BIGFOOT

It's a Fairy Tale...

By Paul Glover BSc.

Grosvenor House
Publishing Limited

All rights reserved
Copyright © Paul Glover, 2023

All rights reserved, including the right of reproduction,
in any form unless written permission is granted by the Publisher,
together with the Author.

The right of Paul Glover to be identified as the author of this
work has been asserted in accordance with Section 78
of the Copyright, Designs and Patents Act 1988

The book cover is copyright to Paul Glover

This book is published by
Grosvenor House Publishing Ltd
Link House
140 The Broadway, Tolworth, Surrey, KT6 7HT.
www.grosvenorhousepublishing.co.uk

This book is sold subject to the conditions that it shall not, by way of
trade or otherwise, be lent, resold, hired out or otherwise circulated
without the author's or publisher's prior consent in any form of
binding or cover other than that in which it is published and
without a similar condition including this condition being
imposed on the subsequent purchaser.

A CIP record for this book
is available from the British Library

Paperback ISBN 978-1-80381-485-8
Hardback ISBN 978-1-80381-486-5
eBook ISBN 978-1-80381-487-2

Acknowledgements

To my wife, I could not have done this without her help and encouragement, she has brought new energy into my life. I love and thank her greatly. To Ellie, my daughter, we all go through challenges in our life, but I want her to know that I will always love her. To Ricky, we went through a lot reaching this point, I could not have achieved this without his time, dedication and commitment. To my friend Pete, I want to thank him for being there during those challenging years when there seemed like there was no hope. To Gordon, for his inspiration and for pushing our understanding of this subject. And lastly, to our forest friends, this book is all about them and without their friendship, this book would never have happened.

Preface

This is my journey, although a few others have joined me along the way at different times, some more than others, yet with special mention to Ricky Bailey, being above all others. I thank them all for being part of this unique adventure.

All of that said, the scope of this book is to give you all a flavour to the path that took me to this incredible outcome. The conclusions that are derived from this are quite simply, far reaching. It took me over 10 years to reach this point, so over those years I have had the personal experiences and the time to reflect and ponder upon everything. You won't be afforded that opportunity, so you will have to make up your own mind here. Basically, science has failed us, and history has been altered to accommodate a different narrative. We have been conditioned over the many generations to accept an idea, yet some of those ideas have been adopted, then altered, from previous beliefs. I found all of this incredible… and given that you need to have an open mind when tackling this subject… you will need to bear this in mind too. I never would have guessed it would take me to this point – everyone I have explained things to have had the same or similar reaction. Disbelief, shock, some have no idea how to deal with this… it is too much – a brain overload and then some choose to ignore. Those who know me well enough or have joined me on this journey have taken time to reflect, review and accept the findings, yet it has still hit them. I know that some who read this book will have difficulty accepting this, indeed, my own older brother will not even talk about the wider subject because his way of thinking is that if it is true, then science would have confirmed it by now. So, that was only in relation to one element…. we never reached past that, let alone the final bombshell being dropped within this book. So, I know the sceptics out there, like my brother, need absolute proof. This, I cannot give within this book, that type of evidence that will persuade the hardened sceptic… but I have left the door wide open for people to take these findings and gain some personal experience for themselves, or to at least dig a bit deeper on what I have offered. Others will take this on board, so I am sure that they will also help confirm everything that I have mentioned here based upon their own experiences and findings.

The consolation I have is the trials that my 19th century hero had to endure in order to gain scientific acceptance – Dr. Gideon Mantell from Lewes, East Sussex. Just like Darwin and Galileo, their claims were not openly accepted. Mantell, himself, made ground-breaking discoveries in the field of palaeontology, specifically related to Dinosaurs. Anyone with a keen interest in his achievements and the difficulties he had to endure are respectfully directed to look in to his life. The 'Dinosaur Hunters' written by Deborah Cadbury gives you a good background to this.

Sometimes I feel like I am following in some of Mantell's footsteps! When the world has a certain viewpoint, yet you are fully aware that there is evidence that points to a different mindset… then it would be considered mission impossible to change that initial idea. Like with Gideon, I hope that my findings are eventually accepted, and this will pave the way to a brave new world. In the 19th. Century it was Dinosaurs! Today? Well, this is my journey, and you are welcome to tag along….

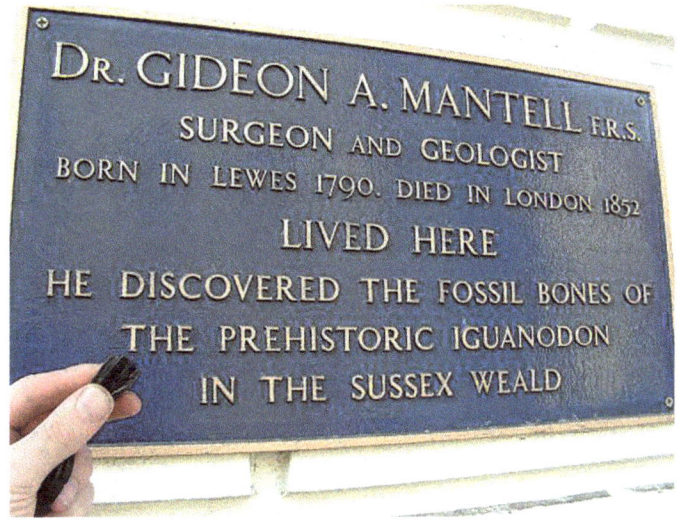

A picture of an Iguanodon tooth that I supplied to the descendants of the Mantell Family in 2007 (on the son, Walter's side, in New Zealand). This is understood to be the first fossil Iguanodon tooth to go back into the family since the 1800's. I made a sentimental visit to Castle Place in Lewes, the former home of Gideon Mantell, to take this picture before being sent to New Zealand.

Close up picture of the Iguanodon tooth. The picture was taken on the footsteps to Castle Place – where Gideon himself would have walked over carrying his fossil specimens all those years ago. Small parts of the root area to the tooth have been retained, so keeping some link to this bit of Dinosaur history.

A Journey of Discovery

Introduction

Where did this journey first start? In answer to this, I just don't know. I initially recall viewing pictures of 'Patty' from the famous Bigfoot encounter from Bluff Creek in October 1967, from an Arthur C Clarke book. My Dad had brought this along with him on holiday to France when I was 14, I remember the chills it sent down my spine thinking about such creatures roaming the planet and it left an indelible mark on my mind at that point. I know the exact location this happened and the feelings that it left with me from that very moment. Little did I know what the future would hold for me back then.

The next memory that I recall was when I was about 19 years old and I had an encounter at an ancient forest in Northamptonshire, Salcey Forest. I was with a girlfriend at the time, we were sat in the car chatting and listening to music with the car positioned along an isolated forest road and it was night time. I turned my head and saw a tall figure (around 6 to 7 feet tall) silhouetted by dim moonlight standing about 20 feet behind the car. You can imagine my shock, so I started the car up, reversed and shone the headlights in the direction the person would have gone and got out of the car and shouted to him. And yet... nothing. No sounds or movement – nothing. I always thought that it was some nosey person – but this didn't sit comfortably with me for all these years. I have no answers... but I sense now that this wasn't a human person whom I saw that night... but it was indeed, a forest person. That could only explain the situation and at that time I had absolutely no knowledge of these types of beings could be living here in the UK – so I only had the one explanation, but one that made no sense.

It was a while before I had any other memorable situations. I graduated in Geological Science at Plymouth University at the time Jurassic Park was hitting the cinemas. This was a great time as the City was also celebrating the 150th anniversary of the term 'Dinosauria' first being used in a Public Address by Richard Owen in 1841. So, there were dinosaurs in the air around Plymouth at this time! This may not come as a surprise, but this led my interest into the dinosaur world and a couple of years later I would find a small section of jaw bone whilst on a fossil hunting holiday in the Isle of Wight. This turned out to be a new species of agialosaur - an early ancestor of the Mosasaurs (the T-Rex's of the Oceans). I went on TV with the BBC Fossil Roadshow (March, 1998) – celebrating a Darwinian Weekend. Because it was a new species, the late Angela Milner said that it would be named after me when the Natural History Museum writes a paper on it. Well, I have yet to hear back what it was named....apparently there is a very long queue of new specimens waiting to be described. All I know, at this time, this was only the 5th example of this type of creature known to science... but since then, many more specimens have been discovered around the world and the last update I had was in 2009 when I was told that the Museum had recently performed a cat scan on it.

10th. March, 2007 – self discovery. On this date I had booked a ticket to attend a ghost hunt at Southsea Castle. This was a castle constructed by Henry VIII to defend the south coast, mainly the important dockyards of Portsmouth, from the French. I was amongst around 50 other people attending and we were split up into smaller groups to go to different locations around the castle to experience different things. It was certainly worth the money as our group had quite a few interesting things happen, and it opened my mind to a different world. At one point we were asked to sit down in a small room next to the courtyard that had a water well, and just meditate for 10 minutes. We were also asked to come up with any names or details that might come to us. Some people called out names at the end of the session... I did too.... and the name I recalled was Lucy and I also came up with an age, which was age 9 (if memory serves me well here). Nothing more was said until we had a wash-up session at the end of the night – about 2 am. The speaker asked who had come up with the name Lucy in the room next to the well. Two of us put our hands up. It was then that the speaker explained the significance. Lucy was the daughter of the Castle lighthouse keeper and she had fallen down the well in the 1800's and had died. She was aged 9 (I believe). The company organising the ghost hunts had previously traced all the historical records to understand the spirits that still dwell within the walls of the castle and Lucy was one of them. Only two people on the night, from all the different groups, got the name Lucy and I was the only person who also got the age correct. So, that night I found out that I did have some hidden abilities, yet little did I know how much of a bearing this would have on me in later life!

My journey to this point revolved around Dinosaurs mainly. These creatures were truly incredible, and they stretched our imagination to the extreme. I was lucky in that my links led me to supply various items to the Natural History Museum, London. The first was the Dino Jaws Exhibition in 2006 where I supplied replica claws and teeth, including the famous Baryonyx claw. The good feedback that I obtained from this led me to supply the Natural History Museum with their summer exhibition called 'Dino Dig' – which featured a full sized Neovenator (a 10 metre long UK predatory dinosaur) skeleton and also an Iguanodon skeleton. These were both mounted in huge sandpits and the public were allowed to 'excavate' their bones. A few years later, 2011, I was fortunate enough to be asked to re-create the bones of Dimorphodon, a famous pterosaur skeleton found by Mary Anning in the 1850's coming from the rocks of Lyme Regis, Dorset. This was for a programme about Flying Monsters and was to be Sky's first 3D documentary hosted by Sir David Attenborough, and for their new 3D channel. I was also asked to re-create Dawinopterus ('Dawin's Wing') as well as to supply other items for the film, which ultimately led me to have three days on set with Sir David himself. And I can assure you that he is indeed as nice as he actually sounds. It was a real pleasure and to be paid for your time, made those few days filming a highlight of my Palaeontological career.

Back to Bigfoot. It was back in 2008 that I ordered a book – 'Sasquatch-Legend meets science' by Prof. Jeff Meldrum. It was originally published in 2007, and this was the first serious look from a scientific stand point as to the compelling evidence into the existence of Sasquatch. Given that I thought Bigfoot, like other cryptids, were most likely mistaken identity (with a slither of possibility), yet I had not delved into the evidence. But here it was... a book that gave me an opinion that they were indeed, theoretically, real! I was hooked at this point – but this was all happening on the other side of the world.... or was it? My reading list expanded, and

Sir David Attenborough looking at the author's replica of 'Darwinopterus' whilst on the film set.

it was Nick Redfern's book – Man Monkey that made me appreciate that there were some weird Wild-man sightings here in the UK too... however, the book was focused upon Cannock Chase in Staffordshire. So, I had to go there, and it was at the planning stage of this that Finding Bigfoot hit TV. At this time, it was gripping watching... tree knocks in the night time forests, weird howls together with eye witness accounts... yet nothing conclusive. And after several seasons... still nothing conclusive... yet the mystery did not go away. That was all going on in America... and in later seasons, the Finding Bigfoot Team dipped into other International locations... Australia, Vietnam, China, Nepal and also.. and much to the Team's dismay - the UK too!

The UK? This set of Isles has been inhabited for thousands of years. Everything to know about the natural landscape of this country is known.... correct? Well, that is what everyone is led to believe, including the team from Finding Bigfoot. They came, they saw and they went home believing no Bigfoot existed in the UK. Maybe in the past, but certainly not now. So, what about these Bigfoot in the past? Well, it is vast... so I will save that for later in the book. However, go to any Medieval church within the UK and I believe most, if not all, will have a representation of them discretely hidden within the architecture, stonework or woodwork.

Although the Medieval period has long passed... the fact that these 'Wildmen' were so revered all those years ago must account for something.... especially in Churches of all places. Even today, there are modern day encounters that have left people scared, confused, or both. One of those modern accounts got me very interested as it appeared to have come from a very credible witness. You can read the witness report online by searching for 'Big Red of Salisbury Plain'. The encounter took place in 2002, but only after several years (2010) that the witness decided to come forward. In summary, a Tank Commander was on an exercise when he (together with his gunner - who only partially saw it) witnessed a 6 foot plus ape like creature run by near their tank. It was running upright, like a human but with a different gait, it had long reddish hair – similar to that of an Orangutan and with a darkish coloured face. The tank commander, George Price, upon reporting his sighting to his superiors, like so many other encounters, was just laughed off and not treated as anything credible.

So, that's the background... my own journey was now to begin!

PART 1

How much do we understand about our forests?

Chapter One

The Search Begins

Where do you start trying to summarise over 10 years of research and the experiences that came from them? Given the exceptional nature to some of these experiences, how will I be able to describe these to the reader? So, with that in mind, my own question to you is – how long is a piece of string? Profound? Now, describe that piece of string.... I am guessing that everyone reading this right now will have their own idea of what that piece of string looks like and how long it is. If I was then to ask what colour it is and what is it made of and how thick it is.... the variations of what people may think of now becomes unlimited. Likewise, I will be trying to describe the encounters I have had and the feelings that I had. To me, these are all personal, but to the reader it will be their own interpretation to that, but I hope my words have done some justice here in getting across what I have been through.

The early days were just like the blind leading the blind. I had no idea what I had started... it was like trying to find the proverbial needle in a haystack, or in this case, within a forest. There was no real information about what I was doing other than experiences that people had in America/Canada and the isolated witness reports in the UK. Did they truly exist in the UK, they appear to have been around in the Middle Ages, but what about now? If they did exist, then they must surely be very endangered, so the population size must be so small, minute! Where would you start looking? Do you follow the same techniques of researchers in the US or do things differently? How would I know what to look for? How do I understand if this is Wild-man activity, is natural, possible other woodland creatures or even human activity? There were so many things that I had to take into consideration... it was going to be a daunting task for sure. Public opinion was also very negative and even hostile to this idea, which I was about to find out when I started my own Youtube channel. The challenge was there, everyone had failed to date in this quest despite vast sums of money being spent on Expeditions, equipment and hours and hours of time – all around the world. What could I possibly offer to this quest? Well, for me, it was personal. I knew that the chances of finding something significant was extremely limited. In fact... impossible, given the odds. I could have done something more worthwhile with my time, rather than wasting it on a lost cause. But I am so glad that I did pursue that 'lost cause'.

I remember the first trip out to a forest on this quest, an ancient forest in Wiltshire. It started off as a bit of fun. I ventured out with my daughter and my then fiancée, now wife. It was summer time, 2012. We walked off the main paths and then ventured into the heart of the forest area. We had watched some of the episodes of Finding Bigfoot (aired summer 2011), so my daughter tried to act out some of this by doing wood knocks and waiting for responses, which, not surprisingly, never came. It was just a bit of fun.... yet the location we ended up at was, by chance, or by selection?, an active area, yet we had no idea about that at that time. In the coming months, I got to call this area site 1. The first of many sites I would get to name.... It was on this day that I noticed loads and loads of sapling tree breaks that I could not explain. Hundreds of them. All done in a unique way. My interest had now been sparked... I would return, but next time alone.

For those that have spent many hours in the forest, particularly ancient woodlands... then you can appreciate how majestic some of the trees are. Some are hundreds of years old... they have stood at that same commanding spot through all the different seasons..... growing from an acorn or seed, punching its way through the forest undergrowth and then reaching for the skies. They are the home to millions of different life forms – insects, birds, mammals, mosses, lichens, fungus, other plants and just maybe... something we have yet to identify. It is always rewarding going on my walks through the forest.... you can sense a different energy that you don't feel at home. The air feels richer... maybe due to more oxygen? but it just feels totally different and maybe this is partly why I would always return... it was therapy as well as a quest for answers.

Sapling tree breaks at site 1, Autumn 2012.

Site one (as it was to be called) was going to become a headache for me. Yes, people do go to the forest, yet nearly everyone sticks to the numerous paths that criss-cross the area. Yes, some will venture off path – either foraging, survivalists, bird watchers etc. I understand that. But what was happening here? Why were so many saplings being broken at waist height and above? All broken in a similar fashion... snapped but not completely broken off. Also, why were there these markings – like nail gouges, seen above and below the breaks? Why were these sapling breaks so different to saplings that deer have used to scratch their antlers on? I had no idea... this was not deer activity, and it did not fit in with any human activity and it was all going on at site one. We are not talking about a few sapling breaks – there were hundreds of them! But was this the only activity going on here? Well, no. It wasn't. This whole process was like learning a new language... a forest language. To the normal person walking through the forest will be blind to all of this. Just like the magic eye books... is what you see all that you can see? Or is there something else hidden from our normal understanding and perspective. In answer to this – yes. There were so many hidden features that looked natural, yet upon closer inspection, it just didn't add up. I was having to train my eyes to adjust to this new forest language. If it's not human behaviour, then what is this pointing towards? I had checked with features found in America that was thought to be associated with Bigfoot behaviours and sure enough, they were being repeated here in Site one. Pinned tree arches, stick structures together with the sapling breaks... however I never saw anything like what I was being faced with in such a small area. To cap it all... here in the UK, we were starting to notice tree branches being stuck in the ground – a term coined as 'ground sticks'. These appeared to be unique to the UK... or so we thought. It now appears that these are being found in the US too. Finding the features was one thing... trying to work out their meaning was another. In summary, I had found various pinned tree arches, several ground sticks, hundreds of sapling tree breaks, 'Y' shaped branches lent up against trees (un-associated with any tree fall), and glyphs. Glyphs are patterns of sticks placed in unique ways; they had meanings - like a notice board. They are usually freshly broken twigs arranged in lines that cross each other. This could not be explained as natural or as normal human

A star shaped glyph from September 2015.

behaviour. There was something going on here and it was all pointing towards Bigfoot like behaviours.... I was slowly falling into that rabbit hole – blindly.

For several months I ventured to site one to record these findings and see if I can identify anything else that had happened more recently. For a relatively small area of forest, this kept me busy. I would spend hours trying to find links or connections to these features, yet to me they all seemed quite random. In the summertime it was tricky to get into site one, as there was thick vegetation at its borders, however using deer paths etc. I made my way into its centre – where luckily, due to the mature tree canopy created less undergrowth. As I learnt more about the site, I tried different experiments to see if I could find who was responsible for the activity here. I invested in some trail cameras – which helped prove a few things.... 1. that nobody (humans) was visiting the area at all for over the three months recording 2. there is a good population of deer that come through the area and 3. some things that I could not explain (like why batteries would fail sometimes).

Picture of a sapling break, and the possible nail markings – Spring 2015, Site 1.

Another sapling break exhibiting nail marking like features.

I kept coming back every couple of weeks to check, so I could see if things changed through the seasons. And it was later, during the winter of 2015, that I saw fresh sapling breaks. How amazing! The area was still very active even during the winter months and why would any humans be coming here to do this?

This site kept me busy for a while and it was during the spring of 2015, I had other ideas about experiments... gifting. I would leave out apples at a certain place – on an old tree stump. I would also arrange a collection of stones in patterns and see if any of them are moved or taken away. I also used jam jars! These were cleaned thoroughly using lens cleaner to remove all oil or residue from the surface and then I filled these with nuts and dried fruit and left these at the gifting area. Why? Well, if the lid was removed then someone with hands had to have done it.... but more importantly... this was to see if it was to get picked up, so I could dust it for fingerprints. And below you can see the results of one of those experiments. The piece of paper held against the glass is a fingerprint from one of my fingers. You can clearly see the difference in size between

Fresh sapling break during the winter of 2014/2015, Site 1.

the dermal ridges of my hand and that on the glass. This was for me a turning point.... it was telling me that this was actual evidence of some beings, other than humans, were here. Maybe, just maybe, I was making contact with the Wildman himself at Site 1!

Before I forget, when I initially came out to site one, by myself, upon thinking there was something odd about the site... and upon entering the forest area from the forest path, I would always bow my head to the forest as a sign of respect. It may seem an odd thing to do.... but this is a tradition I have kept over the last 10 years of study. I like to think that this made a difference.

During the first few years, I had worked alone travelling to the forest searching for evidence. There was plenty of 'potential' evidence to be found. I ventured off to other locations surrounding site one, but clearly more was happening at site one than anywhere else (or so I thought). I found some spectacular tree arches that had clearly been pinned at their tips. I found other tree features that looked odd and some of the videos in

Gifting area at site 1.

Exciting fingerprints that I had difficulty explaining, July 2015.

The author standing next to a pinned tree arch, site 1, Spring 2015.

relation to these can be found on my Youtube channel – 'Anglosasquatch'. I also found a set of tree arches associated with twisted tree branches and sapling breaks. The arrangement was complex and had me guessing exactly how it was made. One of the branches was quite thick – about 4 inches in diameter yet it had been broken with some force in order to make the arch like structure. So, what was going on here? A lot of trouble had gone into making this arch arrangement. I was coming to the understanding that this was clearly not human like behaviours. How could it be? And, why mostly at site one when other areas of the forest was void of activity? Why were my trail cameras not picking anything up? Why no footprints here? It was all odd and yet the more features I found, the more drawn in I was becoming. It was around this time that I found the greatest find yet....

A tree knocking stick: as I just explained, deep within site one I had found an extremely intricate set of tree arches that could not be explained by natural activity or from previous forestry work. It was associated with other activity such as sapling breaks and twisted branches. I came back to this spot on a regular basis looking for any fresh activity plus, this was one of the most impressive features in site one. However, on this occasion, and this was early springtime, within a 10 day period between the end of February and March 2015, I came across a 5 foot long stick that had been left adjacent to the tree arch feature. It was heavy and thick at one end and had been split in the middle due to a pressure crack. I couldn't believe my eyes.... was this stick I was holding a stick used for creating tree knocks? Was this last touched and used by a Wildman? This was certainly a prize to behold! This had been left by this impressive tree arch. It was left there for a reason. I had to understand more what this stick represented. Has anything like this ever been found before? No... nothing, it was a first as far as I could tell. I did a video regarding this on my Youtube channel which explained my way of thinking at the time. And to this date, I still honestly believe this was a stick that had once been used by the Wildman and left at that exact location for some reason. Maybe as a sign or maybe as an offering, maybe it was left for me... there is no way of finding this out. The significance of this find gave me more energy into this project.... I was hooked. Just like in the US and Canada.... they were faced with the same situation.... lots of potential evidence, eyewitness accounts – but no factual evidence that would stand up to scientific evaluation. Could I help add a valuable piece to the Bigfoot jigsaw puzzle? Was this stick beneficial in answering some of the questions?

Pressure fracture to the tree knocking stick. March 2015, Site 1.

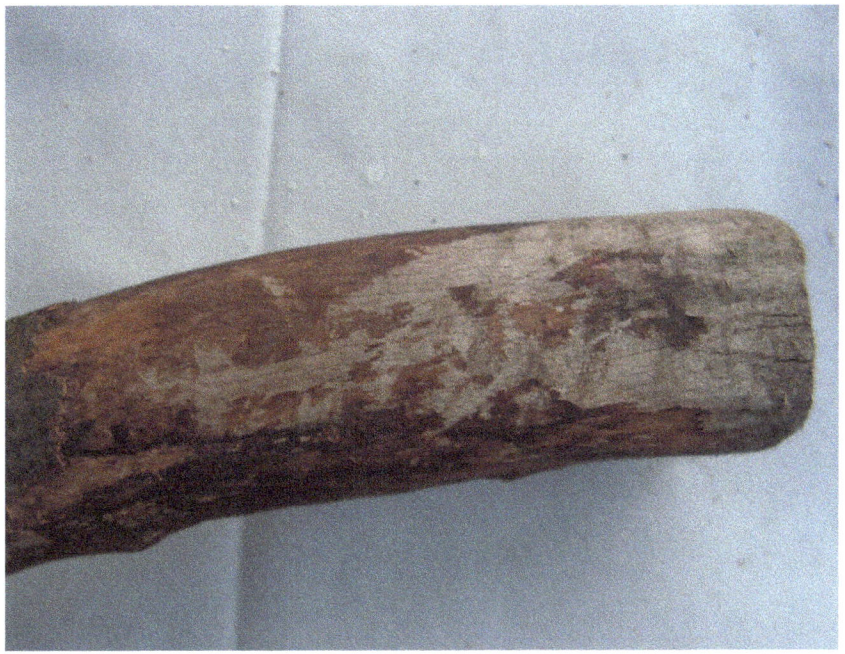

The tree knocking stick showing the wear to the face from repeated use.

My attempts to find that jewel of evidence meant that I had to start widening my scope. Trail cameras were proving no humans were coming to site One, and deer and other woodland creatures showed up nicely, although lesson learnt.... always point your trail camera in a northerly direction so you don't get glare from the sun at sunrise/sunset etc. Sometimes the batteries would fail, and nothing got recorded and this usually involved some recent activity at the site. Strange, yet I never questioned this further as other researchers in the US were faced with a similar situation. I never joined any of the dots at that stage.

Going off path was the best way to find hidden gems of evidence of activity. I ventured from Site One going much deeper into the forest, far away from paths and potential people activity. I was amazed to find fantastic, pinned tree arches, some other tree arch arrangements which clearly were not natural or any hint at human style behaviour. One location, which I called the bottom end of site 1, involved trees that had been twisted around each other – I found two adjacent to each other in connection with a few tree arches and a ground stick. I didn't realise at the time the significance of this, which struck me about 5 years later when I was able to find medieval coins from the 1600's depicting this arrangement and this was associated with the Wildman. Later visits to the location proved that it wasn't just 2 trees twisted like this but 4! The meaning? I still have no idea... but it seemed like it was a gateway into the deeper forest. Please look at the pictures in reference to this. It should also be noted that naturally, trees will always grow towards the light and would not twist around itself as this would defeat the objective. Given this, clear manipulation to these trees has taken place here.

Medieval coin depicting twisted trees in association with the Wildman.

One of four twisted trees that were found at the bottom end of site 1.

Chapter Two

The Search Area Widens

Given the size of the forest and the area that I had covered around Site One, I looked at the map and decided to venture to other areas of the same and connecting forests. There was a lot of ground to cover, and will other areas give more insight as to the picture that was being painted here? Well, the first place I decided to go to was to prove to be the hot spot! So, was it coincidence or was I guided to this area? Just like all the later weird 'coincidences' I feel I was drawn to that area for a reason. The location was to be named site 4. The trees here were generally more mature with both broadleaved trees and a thick clump of massive fir trees. The fir tree area later got its own name as the 'Goldilocks Zone' – which I will explain later about. Site 4 had lots and lots of ground sticks spread over a wide area with concentrations in some parts. The ground sticks varied in size and there were few tree arches – mainly limited due to the maturity of the trees present. I was still going out researching alone and this location was much quieter with very few people visiting, just the occasional dog walker etc.

My initial visits were to scout out the area as much as I could. One area was on the opposite side of a road, yet there was plenty of activity going on there too. I called that area site 5. Here I found an enormous piece of dung, which, for the sake of science, I took home to examine! I have never come across anything quite like it... it was long and thick and appeared to be like a carnivore type poo. I did a video on this on Youtube.... and to this date I still have no idea what created it. Best conclusion was either wild boar or our Wildman? Yet there was no signs of wild boar passing through the area (they usually tear up all the grassy areas) and it didn't look like your typical wild boar scat. Yes, I was dedicated to the cause and had no idea where evidence might come from.

The scat found at site 5. Elongated shape, just like a human or large dog.

The scat found at site 5 being examined. Certainly, all vegetation matter.

Site 5 had a number of features such as ground sticks, but this was nothing in comparison to the features found in site 4 together with the adjoining area, site 6, and it would be here that we would make significant contact. How did it all start? Well, at this stage I was still looking for the relic hominid connection – be it a Neanderthal or other close relative. I was scouting around looking for any signs of footprints – but given the forest vegetation/litter masking the ground, this made it very difficult. Also, given that it was springtime that I had located site 6, the ground was drying out even more. Those footprints never came that year, but one feature did that certainly caught my eye.

Site 6 and the Goldilocks zone was totally different to site 1. There were none of the saplings here as most of the tress were more mature and hence, no sapling breaks to be seen. What was noticed were the volumes of ground sticks. These varied in size from a few inches to 10 to 15 feet high. A few years later I would find the biggest ever ground stick here, associated with a small tree arch and this was the best part of 25 feet in height. Basically, the size of a tree and stuck in the ground! One of the first memories of the Goldilocks zone was entering the area of giant fir trees and noticing a tense head, like a mild headache. I never normally get headaches, so it something that I noted straight away. This 'headache' would soon disappear when I moved away from the Goldilocks zone. Was this a coincidence? Apparently not, as this happened to me repeatedly and it also happened to other people I would take there. A few years later, our first and only night-time session just adjacent to the Goldilocks zone, three of us were all faced with a tense head shortly after we arrived, and it stayed until we all left for the night. My guess was this was some kind of infrasound that was deterring us from spending much time here. I am pleased to say that over the years, the headaches entering the Goldilocks zone faded away, so for me, this was confirmation that I was being accepted in some way, or at least, being tolerated.

So, I came to Site 6 on many occasions by myself during this early part of the year (2015). It always fascinated me as to the volumes of ground sticks here that were obviously not from human activity. The reason for them? Well, that is still part of the mystery. Some parts we believe we understand, but other parts we are still lost as to what they represent. I will try and explain my own feelings about these at a later stage, but for now... this site was special, like nowhere else around the forest. On one occasion I found a pine tree with a shelter like structure at its base. The shelter was protecting from the elements to the north. The tree branches made to create it had all been collected from the same north side of the tree. All the branches on this side, spanning up to 40 feet high had been broken right at the point they join the tree itself. Only on this side and so high up? Not from any other locally surrounding trees... so this just didn't add up. Someone had climbed up that tree and broken all the branches on that facing side and then used them to make a shelter. It wasn't caused by another tree falling along that side – no natural explanation could be seen. So, who caused it? For a human to do this would not make any sense. To climb that high up and 'snap' not 'cut' those branches when more readily available branches could have been used from elsewhere. Some of the branches were a few inches thick, so this also would be an impossible task for a human to do. In my mind, it had to have been a Wildman doing this... but why? I was guessing it was purely as a shelter from the elements... in which case, checking the ground for hair surrounding the tree was a must. Unfortunately, this did not give any potential evidence. Deer and fox hair was found, but no human like hairs.

The first time I found this spot, I really felt like I was being watched. Clearly there was an atmosphere there, but I was alone. I think people can relate to this… you keep looking over your shoulder as you feel you are being watched. But this location was so interesting, I had to study it, yet I didn't want to stay long as that feeling wasn't a particularly good feeling. I did a video at the time and out of the corner of my eye I saw something dart past in the distance. I don't know what it was, but it was large and dark. I guess it could have been a deer… but I have no idea. All I know was that I didn't feel alone there. I did my video then packed up my stuff and got out of there with an uneasy feeling until I reached the car. I had never had that feeling before… even within the Goldilocks zone. It was just weird.

The shelter created by branches all along the north side of the tree, up to 40 feet high, being broken off.

Looking at the inside of the shelter structure. This appeared to offer a very protective layer from the weather.

Chapter Three

4th. June, 2015

It was a lovely summer's day, all the vegetation in the forest was bright green with fresh growth. There was no wind and the sun shone through the branches of the trees creating dappled light patches on the forest floor. It was a lovely day as I ventured back to Site Four. Memories of the being watched feelings were all forgotten, as I checked out the area around the tree shelter for any fresh signs. I headed towards the Goldilocks zone and not far from the tree shelter I found some markings on the ground. These were weird. They were like something had scrapped their hand along the

One of the features found in the ground, prior to casting.

ground, gouging out a trail of around 10 inches. The interesting feature was that this was then partially covered over. Who or what did this? The gouges were too small for a wild boar snout and this only occurred in two places within a few feet of each other. What else could do this? My best guess was fingers (bigger than mine) scrapping the top soil. So, with this in mind, I decided to take a plaster cast of the best one. I got all the gear out, mixed up the plaster and proceeded to make a cast. It was whilst I was doing this that I noted some branch breakages in the distance ahead of me in the Goldilocks zone direction. A deer maybe? But it sounded much heavier, or at least I thought so. So, given that I had to wait for the plaster to dry, I decided to leave a camera rolling looking in that general direction, then I walked away for a while – putting out a trail camera and sound equipment further down the hill, within site Four.

I came back about 20 minutes later to retrieve the plaster cast and my equipment. The cast was still not solid, but I was able to get it out in one piece. I decided it had been a good day so far, so would head back to the

Site of the location where two features were found
in the ground. Also the filming location.

car carrying the cast in my hands in order to prevent it breaking. I walked down the hill towards the tree shelter and walked under a huge mature oak tree. From around 10 feet above my head, I heard an almighty snap of a tree branch and then at an angle, the tree branch landed six feet in front of my feet. There was no wind and the force to break that tree branch, around 2 inches thick and 18 inches long, and given it broke above my head but landed ahead of me.... so my initial thoughts were that 'I was dead!' I had expected that one of these 'wild men' was going to jump out of the tree and kill me. My heart was racing... the adrenaline was flowing.... Do I flight or fight? Given my understandings about Bigfoot, I would not stand a chance either way.... So, I instantly shouted out – and I can remember the exact words to this date... 'I am leaving now... but I will leave you some apples on that tree' and I walked slowly away, listening intently for any further noises and then when I got about 50 feet away, I turned my back to see if I could see anything. Nothing and no further sounds either. I then left some apples on the tree shelter for them as a gift or at least as an apology if I had upset them in any way. I then exited the forest as quickly as I could and letting my heart rate slow down. That day was my absolute confirmation that I was dealing with the UK Wildman. No way on earth was that tree branch thrown at my feet by some type of natural accident or animal. It had scared me... yet they didn't do anything more

Thinking about my own protection, the best I could think of was to make a loud noise like a gun shot.

than throw that branch. I'm sure if they wanted to hit me with it, then it would have done so. From that day forward, I decided I needed some self-protection, so I admit... I ended up ordering a rape alarm and also, as you cannot have any firearms protection in the UK (unless under strict usage/licences) then I got myself a starting pistol. I figured that a loud bang would be enough to scare off any attacks.

Getting back home, after calming down for a while, I decided to review the film footage from the day. On the first quick review I didn't notice anything, but later that night I got the headphones out and watched it in more detail. I was left in shock! There were tree breakages in the distance, there was even sounds recorded that had no reference to any UK bird sounds – in fact, the closest sound I could find was that it sounded like an Orangutang whistle! But the most incredible thing recorded was a figure or figures of upright beings walking in the far distance towards the Goldilocks zone. I was in total shock... there was no forest paths anywhere near that location and this forest area was, to the best of my knowledge, empty of people. The unusual behaviour of the figures moving around was also not like any passing human activity. Then at one point, the most bizarre thing happened. There was a flash of light and then a figure emerged – either from the tree it was next to, or possibly thin air... but the flash of light and its emergence were certainly connected. What was this? I was left totally confused. I had no idea what it meant. I shared the findings with a group of Bigfoot experts at a later date, but nothing conclusive was made regarding this. Yes, it fitted the 'blobsquatch' film footage category as it was a low resolution camera put on long record and was at a distance. I sat on this film footage, yet I wasn't to know just how important this film footage is. It is something, to the best of my knowledge, that was only ever recorded in the famous Patterson Gimlin film from 1967. I didn't know this at the time... but it would prove to be a powerful piece of evidence regarding the abilities of these forest people.

I am still sitting on this film footage as there is still a vicious debate going on about the true nature to Bigfoot. For me, there is no debate, I already know the answer to this, but at this stage in 2015 I was still firmly sitting in the biological primate camp. So, I was very confused, and it would take night-time adventures that would finally turn my head in a totally different direction.

June 4th. 2015. Film recording. Left hand side, half way up, you can just make out a figure in the background.

Close up image of the figure in the distance towards the Goldilocks Zone. It appears to have a high shin angle and a uniform blackish colour.

Chapter Four

First Night Time Encounter

To this point, it was just me doing solo visits to site1, 2, 3, 4, 5 and now 6. I had started my own YouTube channel and was fending off all the negative and hostile comments being left for me even daring to think we had any Bigfoot in the UK. Yet some people had an open mind and through contact with one of those individuals that I decided to get someone else on board. That person was very much a countryside man and was aware of some strange things happening in the UK, especially around Cannock Chase in Staffordshire. Well, I introduced him to Site 5 & 6 and he was stumped as to what was going on there. He could explain a lot of the natural things, but others, he was lost for explanations. So, I was glad that my own rationale way of thinking was also standing up to agreement here, yet this only explained the weirdness and not who was behind it all. It would take night-time visits in order to test our theories… but when?

It was around this time that I made contact with a good friend, Ricky, who we had not been in touch with for a couple of years. I still remember the look on his face at the Pub as I was trying to explain to him that I thought that we had Bigfoot in the UK. He remained calm, polite, understanding, and he kept his personal opinions to himself that night. He said he would need convincing, so I offered to take him out to the forest and take a look for himself. It was 6 months later that Ricky eventually told me that based upon the conversations we had that night, he classed me a nut job! Now, he feels bad for ever doubting me. This is our story regarding our first night-time encounter.

As promised, I took him out to the forest for the first time to search for evidence to back up my claims. We went to site 4 and Ricky was faced with multiple ground sticks, tree arches and weird stick collections. He had no answers for this but was trying to put this down to human behaviour…yet it made no sense to him. It was at one point where we found a pushed over tree with another tree pushed under its roots that made us stop and ponder on what was the cause. I had already warned Ricky of the rancid smells that come from no-where before we arrived and yes, we were hit by an extremely strong rotting meat/rancid smell at that point. It came from nowhere and was potent. Ricky's reaction was… OK, there is something not quite right about all of this and he was now interested in knowing more. He was not yet convinced it was Bigfoot, but he certainly had no answers regarding what it was.

The author next to the nest like structure found at Site 6, standing next to its entrance. You needed to crawl through a small space in order to find a large interior. This could easily fit three adults lying down.

So, July 3rd, 2015 we decided that we would just sit out at a location near to a nest like structure that we had found at Site 6. It didn't get dark until very late and we just sat on the ground listening to everything. At one point we heard a litter of wild boar piglets in the distance, so we were on our guard should they come near to us. We heard a couple of knocks in the distance, but certainly we could not relate that to Bigfoot – it could have been anything. So, our first night was relatively quiet, uneventful and easy going. It helped us gain a bit of confidence for future trips there.

It was July 9th that we got our 100% confirmation. The location at site 6 was a good distance off any woodland paths. We didn't see anyone at all, we had the forest completely to ourselves – or that's what we initially thought. I had started to get hold of some spy equipment – which was not movement activated, but was noise activated. I set one of these up around the nest like structure. For our defence, as we know they don't like trail cameras… I put one of these directly behind our backs (in case anything tried to approach us from behind). Ahead of us, about 200 feet ahead, we used Ricky's digital camera on a tripod and put it on long record. It was dusk and started to get dark, so we left it recording and then returned to our night-time spot and just listened to the noises from the forest. I had a bionic ear that we used to listen to things surrounding us and it was about 15 minutes from when we sat down that we started hearing tree breaks and heavy footsteps directly ahead of us. This was not possible. Was this really happening? Directly ahead of us was just the forest, no paths etc. There were no lights, should it be some random human etc. We were using the bionic ear so we could hear solid, heavy bipedal footsteps crunching on the forest floor. This was incredible…. There must be some truth to this myth! Our hearts were thumping, the adrenaline was rushing… and this carried on for about 10 to 15 minutes and then, in the silent forest, three almighty tree knocks! Ricky said that was enough for him… he was convinced 100%. There are Bigfoot here in the UK and we just had our first close encounter with them. I was no longer the nut job.

We didn't need to go to the USA for an encounter with Bigfoot. We have them here in the UK. Just within a 30 minute drive from home. We were now faced with the same problem as they have in the USA, how do you prove it? Well, we weren't going to be doing it that night. We chickened out. I wanted to stay longer, but Ricky was really scared, I was scared too, as they were literally just a hundred or so feet ahead of us. How little did I know back then… but this was day 1 of our first real encounter and 100% personal confirmation. After those three, crystal clear, loud wood-knocks our torches went on and the sounds disappeared. We cautiously collected our gear, and I will be honest, it was real scary collecting the camera from the direction we heard those bipedal steps. We got back home, grateful that we had survived, and then listened to the recording. Yes! We had recorded one of the three wood-knocks! This recording is available on Youtube for you to listen to. It was our proof, and also real evidence of their existence here in the UK. I would recommend using earphones to listen to the recording and do it in a silent, dark room. You will then be able to hear what we heard that night. That first wood-knock will have the same effect on you as it did on us! When I added this to Youtube I had a Bigfoot Researcher from Alaska listen to it with professional sound equipment and she said that it was genuine Bigfoot sounds that we had recorded. She was very impressed with it stating: 'This is a really remarkable recording' and her confirmation was: 'Bic clicks'. I have never heard of these before, but she explained it was hidden within the recoding and is associated with other Bigfoot recordings in North America. Not only that, she said that she could faintly hear some sort of language being spoken. Now these sounds, no matter how many times I listen to it, are a mystery to me, but she could hear them. So, here was our first bit of evidence and exceptional piece of evidence too! For me, this was proof that we were dealing with the same entity as they were in North America.

Penotia Sesquai 5 years ago (edited)
Hi as - this was a fascinating night sit! My heart was going boompa-boompa-boompa right along with yours. Quite the journey!! So, I heard some things: I heard the periodic rustling, and vokes even when it was distant 00:27 (f), 00:53 (f/y), 01:08 (?), 01:57 (m), 01:58 (f/y) and so on. There was an interesting male chant-like communication at 02:33 and more vokes at 02:44 (m), 03:20 (m/y) and so on. I'm not so sure that was a squirrel 03:54 +, or a bird 07:28. The vokes stopped when he/she/they were at their closest, or in their greatest danger. I also think they had tacit permission to do this experiment as the "whack" was the all clear signal. The vokes started up again when they were again away from your position. There is a beautiful songlike voke (f) at 08:00 that goes on for 8 or 9 seconds. This is a really remarkable recording. Thank you so much! p

If you have exceptional hearing or expensive sound equipment
to review the recording, then here are the references to the vocalisations.

But what about that spy camera that was left just a hundred feet away from us at the nest? Upon listening to it, I was in shock! There was another Bigfoot approaching us that night on our right-hand side. The spy camera was tiny, the size of a lighter, and I hid it on a tree branch under some moss.... Yet a Bigfoot had approached it. The recording had started with footsteps approaching it, and then something had touched the branch with the camera on, even sniffing it, and then the recording was 'zapped' (a term used where Bigfoot are believed to use infrasound, or some other technique to effect people or equipment). This recording has also been added to my Youtube channel. We now know that they had approached us on at least two different sides that night and only a hundred feet away from us. That was enough to give you nightmares. Will we have the stomach now to go back and continue with the night-time encounters?

Definitely – Yes!

Chapter Five

The Summer of Night Time Encounters

10th. July, 2015

'Given the sensitive nature of this subject matter and following the close encounter as of last night, plus the previous week, (name withheld) and myself have decided that we go quiet on this as we are getting close to actual contact with these forest friends. With this in mind I have called this group Project Zulu. Please ensure that this group stays in the dark. This will be our sounding board for further research as time continues. A select number of people will be invited to join the group over the coming weeks'.

Due to the recent activity that we were experiencing, I decided to reach out to well-known researchers in the Bigfoot community. This was in order to share my information and my thoughts on this, plus as a mechanism to track my own activity, should I ever need to come back to it later. I am so glad that I did this, as it has offered me a wealth of information that I will use within this book. Various entries from this have been copied across below (with only minor alteration for spelling etc.). I called the initial group as Project Zulu and this was a private group saved on Facebook. The above is the first recorded entry. The reason I called it Zulu was, and I quote from the group:

> 'Stop throwing those bloody sticks! (in a Michael Caine accent). I put this as a result of the stick that was thrown at me, but the real reason why the group is called Project Zulu is a little more profound. Clearly the Zulu nation was a collection of tribes (like Bigfoot are). These tribes were considered uncivilized and backward by

the British Empire at the time and was hardly respected. Settlers encroaching on their land pushed the two sides into confrontation. The British Empire, with all its modern technologies underestimated these tribes and their arrogance was given a beating at the slopes of Isandlwana. There are certain analogies that can be drawn from this - dare we care to ponder on that thought? Project Zulu.'

It's ironic how those words now resonate 7 years later.

It was also at this same time, having had that initial night-time experience, that I decided to invest in a thermal camera. These were very expensive, and, bulky back then, so I was fortunate enough to get my hands on an ex-fire service thermal camera. It was heavy and the battery didn't last very long, but the night vision effect was really good and it was certainly a start, and we thought this would be heading in the right direction. If only it had the ability to capture images, then this would have been amazing. So this is a record of our next night time encounter – directly copied across from Project Zulu and was written by Ricky.

Thursday, 16th. July, 2015

'Attached is an image I have drawn depicting what I can only describe as a humanoid figure seen through a thermal imagery camera during an exercise the night of Thursday 16th July 2015 at site 6.

We began the night with high hopes on the back of last Thursdays session so repeating what has worked for us already we set up cameras during dusk and as the sun set we let darkness take over the forest as we sat and waited in silence. The night began very quiet, little to no activity for the first hour or so as we sat in silence. After around 90 minutes we began to think about packing up and calling it a night until at around 22:40 we suddenly heard an almighty rustle in the distance around 60ft-70ft away from where we were sat. Grabbing the thermal camera and focusing on the direction of the noise I eventually caught the image I had drew. A hominoid figure moving from right to left, rather sluggish through the wood. It had broad shoulders a rather oval shaped head with a warmer thermal signature around its face, not so much the body. I'm assuming it was walking side on but slightly with its back to us as though it was walking away at an angle. In the excitement I passed the camera to Paul Glover who tried again to focus in on what I had seen but the figure had soon disappeared further into the distance. I never saw any legs only the head, shoulders and top half of the torso. I drew in a bush as the only explanation to not seeing the legs, but the bush could not be seen in the thermal energy camera, only the trees.

The only sceptical explanation to this would of course be the figure being a person. Now at the time, Paul will agree, we could barely see our hand in front of our face, the forest was pitch dark. For a human to be walking around right next to us, without a torch, in such a remote part of the forest as site 6 at this time is simply incomprehensible. This could only be a bigfoot.

After the events, the forest returned to silence and after waiting a further half hour we decided to call it a night. We switched on all our torches and aimed them at the direction of the figure and it was only then that we had come to the conclusion that the figure was in fact behind either the nest of site 6, or a nearby bush slightly further back from the nest to its right. We will be investigating and re-enacting this event in the daylight at some point so that we can get some more facts and visualisation on the sighting. In conclusion a very exciting night indeed.'

The thermal hit that was encountered on 16th. July, 2015.

Close to the night-time location from that night. You can see the nest structure in the background and the being must have been standing behind it, so estimate being in excess of 8 feet in height.

Now we had actually seen one of these forest beings within a hundred feet of us – this was getting crazy! They were just as interested in us as we were with them. For the record, we did nothing to encourage them to come to us – we just sat silently in the dark and allowed them to come to us, and it was working!

11th. August, 2015

The excitement and heart racing experiences that we had that summer was dragging us back repeatedly. With the idea that we were clearly dealing with a clan of active Bigfoot beings here, close to home, I decided I would improve our chances of getting some night-time footage, so I bit the bullet and obtained a second hand Flir C2 camera and also a cheap Chinese thermal camera. We took these out with us on our next night time visit – August 11th. 2015. Again, this is a copy from Project Zulu.

'Awesome start to our night time investigation last night... 9 crystal clear wood knocks just a few hundred metres away from us... I think we must have disturbed them when we were walking to go set the cameras for the night. I'm sure they saw us and warned the rest of their group. The wood-knocks happened in quick succession... just a matter of seconds apart from each other. We have got 4 recorded on camera.. but you will need headphones to hear them above the walking/talking... with all the excitement Ricky counted 8, but actually it was 9. We now know that they are living close by as the direction of the wood-knocks were coming from the same area again and this was at dusk. I believe they are living in site 4 somewhere... where I had the tree branch thrown at me previously.'

There is a recording of this encounter on Youtube – British Bigfoot – 11th August 2015 – 9 wood-knocks at site 6.

12th. August, 2015

Given that we were now fully aware of the Bigfoot like creatures were here, living alongside us in the UK and they were very similar to these forest beings in North America, I was starting to think of names we could call them. Woodwose was the medieval name used for them, but we were now in the 21st Century. Is there something more appropriate we could use, or at least I could use, to name them? So, on this day I came up with the name 'Anglosasquatch', and here was my reasoning for it:

> 'I am proposing the name 'Anglo-sasquatch' or for short, Anglosquatch for the forest people that we are currently studying here. The reasons for this are threefold. Firstly, in recognition of the word "Sasquatch" which is an Anglo word created by a Canadian using various First Nations' words for the hairy forest giants, so this is in direct connection and with absolute respect for the popular name for these forest peoples. The name woodwose (or its original Anglo-saxon name – 'wuduwasa') – meaning hairy man of the woods, being the old medieval name for these creatures, so combining the Anglo pre-fix, is an attempt to recognise the popular name for these creatures and giving it the Anglo prefix (both for location and historical reference) to bring the name into the 21st century and linking it to the valuable work going on across the pond in pursuit of the same (or similar) creature. I hope this may one day get

adopted, but of course this is just my proposal, but I will be referring to the creatures in our study area as Anglosquatches from now on.'

During this summer I was very active researching this forest location. I was doing all sorts, collecting hair samples wherever I could, leaving out gifting presents (with glass jars), filming whatever features I could find, leaving out long recording equipment, spy cameras, and all sorts. All this activity was recorded on Project Zulu, but there is little value in including all of this in this book, as this would just turn a single book into a volume of books. So, I will just highlight the major points that we came across leading up to the major bombshells being raised later in this book.

The Autumn of 2015 proved to be the start of the weird experiences. We were seeing light anomalies in the forest at night-time. And it was on the 4th. Of October, during the daytime, when I was visiting the Goldilocks zone that I witnessed a very strange event. It only lasted a few seconds, so not enough time to take any pictures, but it was very clear to me – at about 200 feet away. I took a picture of the area immediately afterwards and then at a later date I tried to do a photoshop recreation of what I saw. This is shown below.

4th. of October, 2015 – light anomaly at the Goldilocks zone (re-creation).

You can see from the picture that it was a bright glowing red light with a diffuse glow surrounding it. It was about 3 feet off the ground, and it just appeared out of nowhere. This was all getting weird now.

In September, I had also cleared out three bags of leaf litter from within the nest like structure in an attempt to find any hair. I retrieved around 500 individual hairs and two of these proved to be very interesting. Here is an account that I wrote on Zulu surrounding this:

> '26 cm long (10 inches or so). This is the one I am very interested in. A closer look here… it looks human like… but certainly appears to have some differences. As can be seen, the colouration is a reddish brown…. But looks darker away from strong light. A few images were left out of the last album that relate to the longest hair. I don't want to mess around with the long hair using any chemicals on it in order to obtain scaling patterns… not until I know it is safe to do so… so these are just images of the actual hair under the microscope… Looks cut… but this is from the root end of the hair… I have experimented with my own hair why this could be… and I can now relate to this and the hair follicle could have been snapped.
>
> Decided to use the SMALLER piece of hair to do some microscope work with it to deduce if it could be human or otherwise. Alas, in the process of using PVA glue to make hair scaling casts it broke in two… I usually use another medium but did not want any chemical reaction with the hair, so used pva glue instead which is water based. First time I have used this, and I think the glue was a little too thick, so I didn't get any really good scaling patterns and unfortunately it got broke the hair in two removing it. This said… I am pleased to say that so far the outcome is looking good. The hair looks like human hair, but clearly it shows signs that it has not been cut. The hair colour looks a reddish dark brown (similar to local sighting reports). There appears to be no medulla (QUOTE: Hair presumptive for Sasquatch looked human but did not have a medulla). The length of the hair removes the possibility of being a woodland animal hair. The hair scaling is an imbricate style (human like) and they seem to be, generally, wider than my own hair scaling patterns (which is what I have seen before, but could also be related to the position along the length of the hair). I will

do a short piece looking at my own hair and show a few things that I noted.... which help confirm some of the feelings related to the hair recovered. So far, so good....

Feeling a little more excited about the hair sample gathered now after doing some extra reading.... only 1-2% of the human population actually have red hair and from this only 2-6% in northern and western Europe. What would the chances be that it was a red head going into the nest like structure and being the only human like hair found in it? Less than a 10% chance it now seems.... Eyewitness accounts for these beings is that they were reddish brown in colour.... clear daylight sighting 15 miles to the south of the research location.'

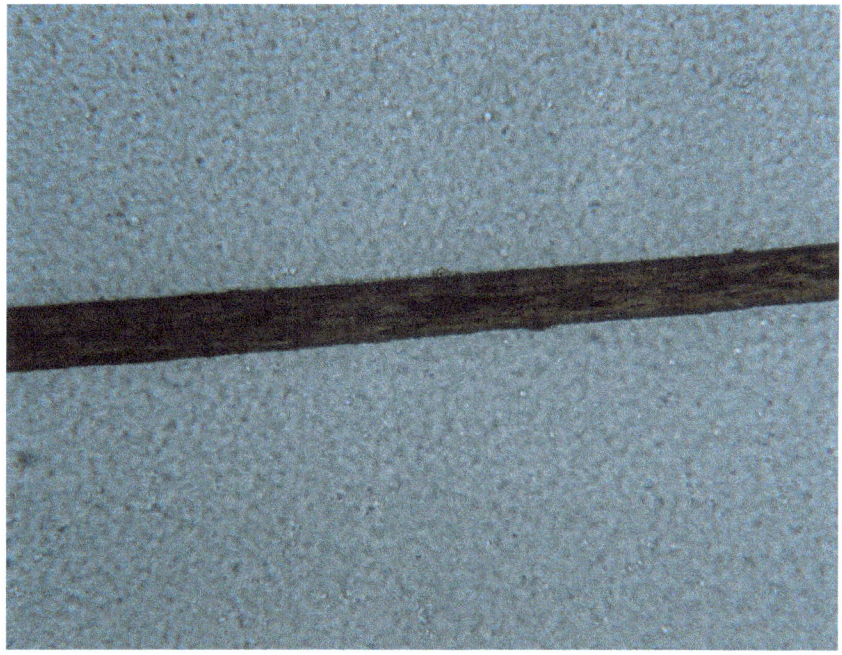

Hair sample with a fragmented or no actual medulla.

Note: I want to make it clear that I am no expert on hair analysis, but I was using my own detective abilities to look at hair samples as best as I could. Hair was considered a good avenue for scientific evidence supporting their existence.

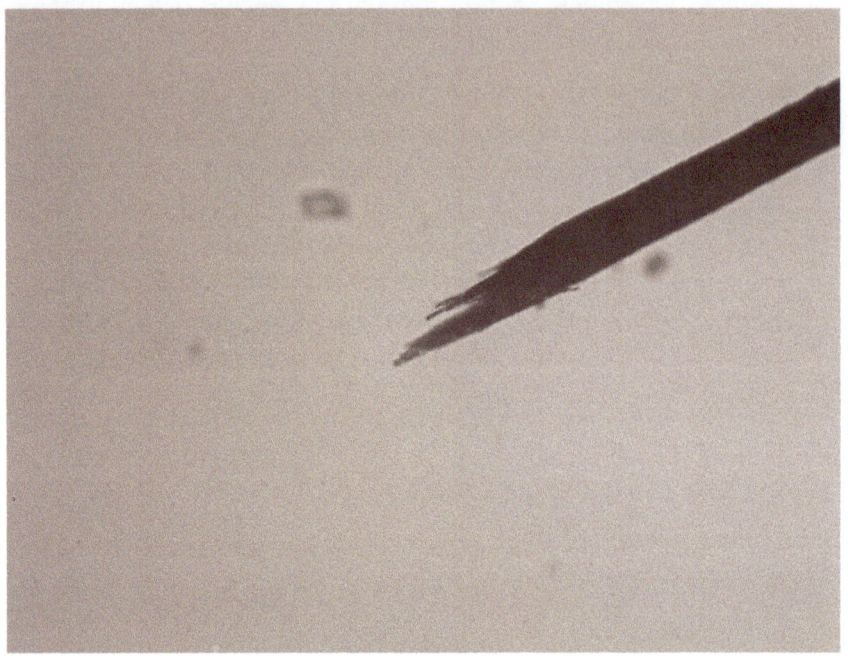

Hair sample indicating un-cut/split hair ends.

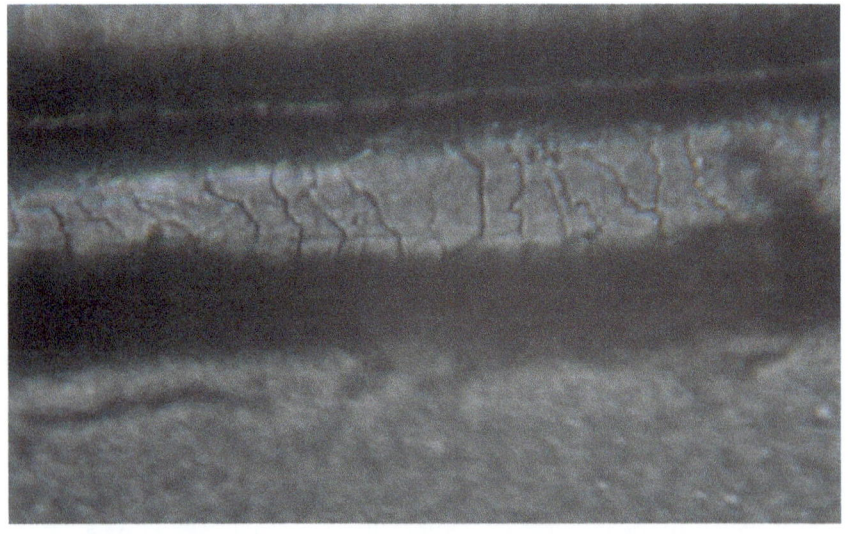

Imbricate hair scaling pattern with wider scaling than my own hair.

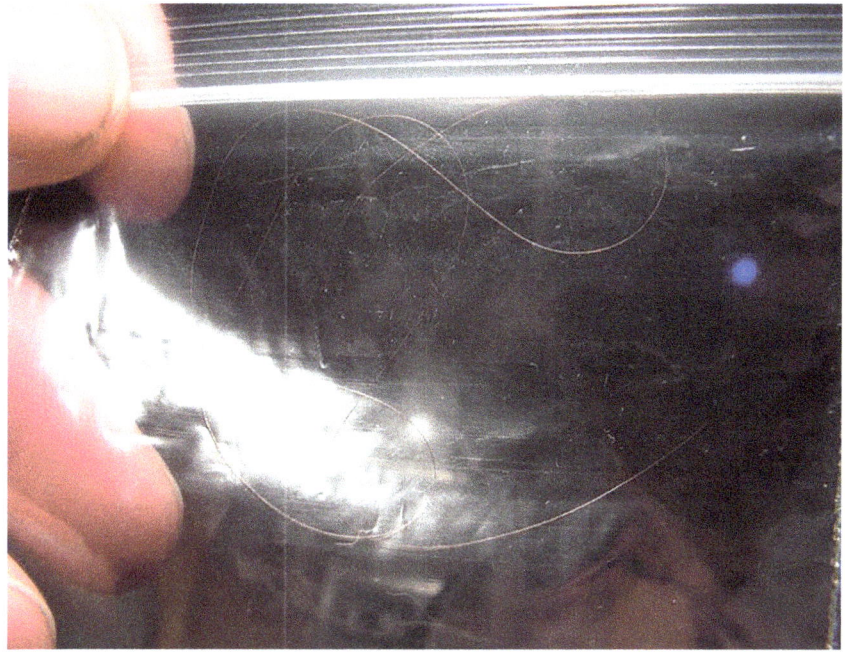

Picture of the 10 inch red hair collected from inside the nest like structure.

No further news on the hair sample as it was sent to the USA for further study.

8th. October, 2015

'WOW! WOW! WOW! WOW! Just come back from site 6, bugged out early... mostly as a sign of respect for the forest people.... but what an amazing half an hour of a night investigation! I mean.. within 2 minutes of setting up for the night I had a tree pushed over!!!! Got it recorded as well... now, for the record, the night was very calm, no wind and I was by the car by the time the tree got pushed over. It all happened very quickly before I had chance to set up... just got one microphone out - the parabolic one (alas, pointing in the wrong direction), but it did record some twig snaps at the time of the tree being pushed over.... so, there is only one answer to this. After the tree was pushed over I was listening with the parabolic mic. and heard loads of heavy twig snaps. I will listen to the whole of the recording first... (which I know has recorded the tree being pushed over) and see if I can pick out any other sounds.'

This was the entry from Project Zulu regarding a solo night-time encounter that I had at the bottom end of site 4 where I park the car. I was starting to be brave in doing night-time research by myself, but keeping close to my car. I was rewarded literally seconds after putting the recording equipment out with a tree being pushed over. As you can tell from my reaction, I was overwhelmed, but I can assure you, at the time I was freaked out.

This is the tree that I believe was pushed over. It had strong roots that had clearly been snapped, and there was no wind that evening.

26th. November, 2015

'Back from the forest... what started off as a very depressing start to tonight's investigation turned into something to really smile about. Basically, the forestry work had started (and we believe now finished) in part of site 4. The important areas look like they have been unaffected.... BUT... what made us really smile was the FACT that our Forest Friends had started to retake their territory.... fresh ground sticks had been stuck in the ground close to, around and in the deforested area. They had retaken their land! We also found a small deer skull cap placed on a small

2 pictures: 26th. November, 2015. Deer skull cap
found in a small tree surrounded by massive ground sticks.

tree - it was not a bird or a human doing this for sure... why? Two massive grounds sticks placed right next to it... this was not here before the forestry work... so maybe a sign! Anyway... will give a more of a full report tomorrow with pictures... Activity? Bright white flash of light from Site 4 around where I got the film footage in the summer. Contact? Just lots of whistles.... initially we thought a bird... but after around 10 plus we knew this was no normal bird sounds at night and believe it was them. The night was PERFECT.... the bright moon gave us awesome visibility... and the air was so still we could hear everything going on. Twig snaps heard... but in the very distance... A very good night after what we initially thought would be a disaster for our future trips there...'

December 3rd. 2015. A long recording spy camera that I created by placing it into a small log. You can just about to see the camera lens. Surely this will outsmart them.

As you can see above, I was trying all sorts of methods in order obtain evidence and footage. It just didn't work the way I wanted it to. Yes, there were some recordings, but they were usually followed by electrical interference. It also occurred to me that they were not happy with me doing this as when I removed a long recording microphone away from our night-time location, I found upon our return that the stump that I had used to hide the microphone had been kicked in. There was also a lot of weird stuff happening now, so were we going about this the wrong way?

Chapter Six

A New Year and New Experiences

Again, I will just copy the extracts from Project Zulu where interesting events were to happen. The early part of 2016 was quite flat. There had been forestry work around site 4 which we didn't appreciate, and I'm sure that the forest beings didn't either. I would go out by myself during the daytime and look for recent activity and it was rewarding that there was plenty of it where the forestry work had been. Fresh ground-sticks everywhere as if they were taking back their territory, so a good sign that they were still very active around here. There were not as many night-time visits as it was not pleasant sitting out in freezing temperatures at night.

21st. April, 2016

I was accompanied with my wife and another person to do a night-time sit at our usual spot at site 6. It was a dark night with just small amount of moonlight. Basically, it was a fairly flat night as far as actual activity goes, but three of us witnessed up to three flashes of lights within the canopy of the trees. It wasn't just the flashes of white light, lasting a second or two but there was also the heightened feeling of 'energy' surrounding us. A short video of this was added to Youtube – 'British Bigfoot – our resident sceptics view on things.'

20th. April, 2016

My feelings at this point and further consideration of the film footage I took on 4th. June, 2015.

'OK, we are reaching the point now where we have to start thinking outside the box and taking a look back upon our experience and understanding to what has been going on here at site 4 & 6 over the last 12 months. Points to note:

1. Bright flashes of light seen in the forest at night not associated with human activity.
2. Cameras being interfered with or just failing
3. The ability to be real close to us (enough to smell their pungent smells and yet not see them) or appear on thermal cameras despite being so close.
4. The glowing red light I saw in the Goldilocks zone during the daytime last October.
5. No footprints found to date? Despite knowing where they have passed by.
6. Tense head feeling near to the Goldilocks zone from time to time.
7. Few eye-witness accounts.
8. The following video clip.

Given the above, I think it was about time to look back on the film footage from June 4th., 2015. What to me first appeared to be some sort of an anomaly is now turning into something of crucial importance I feel. M K Davis did a video (link shown below ages ago (which has always been at the back of my mind)) about the famous Patterson Gimlin film and a bright flash of light and then the emergence of a possible second Bigfoot at the end of the film. I feel this has also been captured here too. Alas, the film quality isn't great... plus given it is at a distance... but clearly there can be seen a bright flash of light and then the emergence of a figure. I have made three short videos focusing on this element - the bright flash of light appears at 12:53 in the original film. Alas, the quality gets poorer the closer the video is cropped.

I am now of the opinion that we are not dealing with some relic hominid, the likes of which we have been searching for and the time has now come to start thinking outside of the box as to the true nature of these forest people. I have no conclusions whatsoever, but my focus now will be tailored to keeping an open mind that we might be dealing with the absolute unknown here, yet these forest people are certainly real.

If we are now dealing with quantum physics here, then this is outside of my remit now!

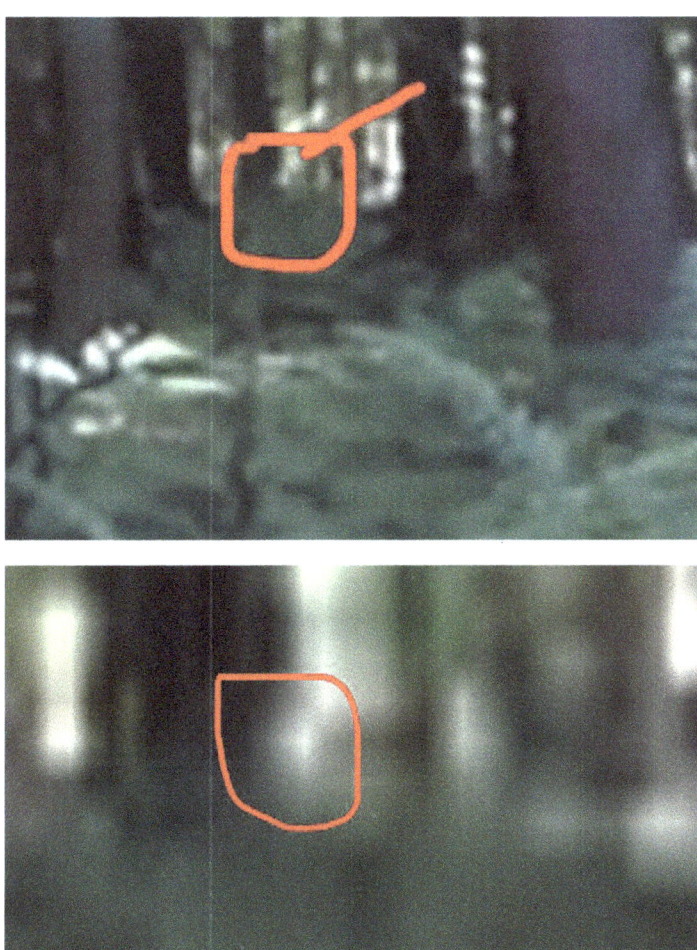

You can see, faintly, the light source coming from next to a tree and the emergence of a figure following this. This is not well shown here (as both pictures are cropped and enlarged), but in the video clip you can more easily see the flash of light and a figure following it.

Below are the details that I would like you all to review on Youtube – a video created by M K Davis looking at such an anomaly on the famous Patterson Gimlin film. If we are ready to accept that the film footage is genuine, then we need to accept that this light source and potential second figure seen in the film is genuine too. Remember that Roger Patterson said that they understood that other Bigfoot were there at the time. So, if this is the case, then this adds further credibility to the footage here recorded in the UK. Because of all this controversy, which M K Davis confirmed he was targeted as a result, then this is why I have sat on this footage for so long. The whole environment needed to change before I would consider releasing this publicly.

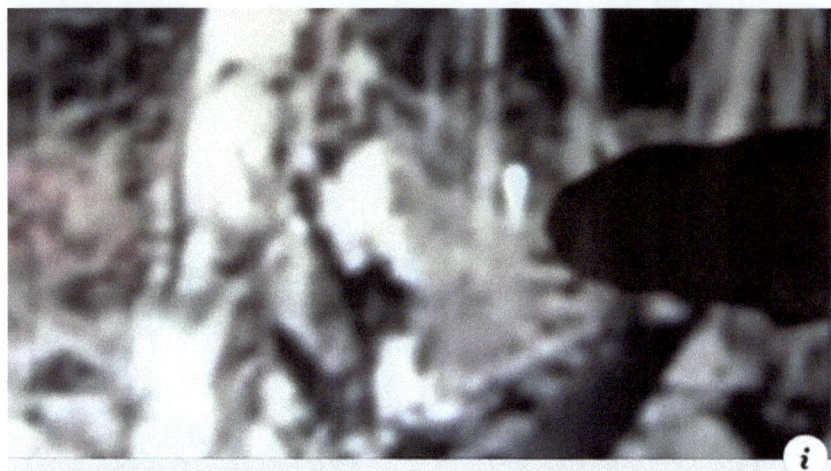

M.K.Davis discusses a possible second Sasquatch in the Patterson film and other anomalies.

M K Davis does, at the time, a controversial video on a light source seen in the famous Patterson Gimlin film. Not so controversial now.

14th. June, 2016

This is a report from our first and last ever night-time visit to the Goldilocks zone:

> 'Reporting back from our night time investigation of the 14th. June next to the Goldilocks zone. This is the first time that we can confirm that in the presence of 3 different people, we all experienced the same feelings. I have seen this many times before when you get close to this area, but this evening it was very noticeable, so I asked the others and we all confirmed the tense/headache like feeling just at this location. After about an hour of sitting there, it went... and none of us had that feeling before we arrived. What does this relate to? Well, it does confirm that there was a change on a physical level that

we all picked up upon. This also confirms that there is something very unique about the Goldilocks zone that we have singled out, as this is generally where we see the light anomalies. Alas, we didn't stay here long enough this particular night as Gem was feeling tired and was getting cold. That said, around midnight... we did start to experience twig snaps near by - which we usually attribute to their presence - however, this is also the point in time that Gem wanted to leave... so we dipped out at this moment... No lights seen that night and the only experience we had was a creature running off near by us making a terrible guttural roar like sound... We were of the opinion it sounded just like a gorilla charge roar... as it was nothing like a muntjac alarm. However, I had the bionic ear on facing the wrong way.... so could not be sure. On review and checking it against other videos... I feel it is the alarm sound of a roe deer... although it really did sound like a gorilla too. But, given this happened early in the evening, I feel it is more likely to be a roe deer. Next time, I would like to stick it out next to the Goldilocks zone.'

We never came back at night as we believed they didn't want us there, so we respected their wishes.

14th. July, 2016

I had brought on board another keen researcher that I had made contact with through Facebook and I took him around Site 4 and 6 and he was blown away by the number of ground sticks seen. I was explaining things to him when we came across a small tree arch with a ground stick stuck in the ground directly next to it. It was on a deer pathway leading away from the Goldilocks zone. It was upon inspecting the tree arch that I noticed a bent twig with some of the leaves attached and evidence that other leaves had been bitten off. I picked this up for further analysis. Well, it was later when driving the car back, going past part of the Goldilocks zone, that I stopped just to explain things about that area and say my farewell respects. This is the report from that day:

'The weirdness of the Goldilocks zone continues.... this time, just as we were leaving, I said we should just pay our respects before we left (like I normally do when leaving the forest), but as we were driving past the Goldilocks zone, I pulled the car over and say we would do it here. All was well until we tried to drive off - the car would not start! This is the first time this has ever happened and it turns out, after getting the garage to look at it - that is was a blown fuse in the engine compartment. The timing was too much to have been a pure coincidence. I had driven the car from the normal parking place to a position half way up the hill (as I forget to bring back some interesting evidence - more to follow on that later on), so left the car there and dashed off into site 6 to collect it where I had accidently left it. Sure enough the car started fine there... but just outside the Goldilocks zone it failed. Now... what sort of sign would you read into that one? We ended up having to jump start the car to get home.'

I was to later find out that this was a burnt out alternator. How did this happen? No warning or anything... coincidence or not? I would argue that there was no coincidence for this happening at this point in time.

15th. July, 2015

'Ok, here are some close up pictures of the branch in question. Its from a Goat Willow tree - the leaves are edible to humans and as well as food, they also have medicinal purposes. The branch had been ripped off a tree and CARRIED to this location which was right next to the tree arch and a ground stick along one of their believed paths through the wood from the Goldilocks zone. The leaves have been stripped at the end of the branch (all fresh growth) and other leaves lower down were nibbled etc. Can any features to a bite pattern be determined here... I will look into some of these individual leaves in more detail. I think we can discount deer and humans behind this being left in the manner that it was - so, my own feeling is that this was some of their food from the previous night - given how fresh the leaves were. The search continues... but a good find.'

'Yes, if we had a Scientific team behind us.... I think a saliva/DNA swab would have been a worthwhile check.... Oh well, we need to work out from these remaining leaves if it was a human like bite mark – so you can guess what experiments I'll be doing over the weekend. Will check the remaining partial leaves for bite indentations on the surface of the leaves too. All fun eh? But I honestly feel this was some of their dinner from the previous night.'

The goat willow branch discarded next to the small tree arch.
Notice the fresh leaves have been eaten but older leaves have been left.

Shape of the bite markings on one of the older
leaves. The fresh new leaves had all been stripped off and eaten.

I did a Youtube video on this branch, and showed extra information that I was able to find together with my own experiments looking at the bite patterns produced. I feel this is something that they had eaten and discarded the branch next to one of their features… maybe, just maybe, as they knew that I would find it there later on. The other interesting observation was that only the leaves at the end of the branch had leaves stripped/eaten away. It never occurred to me until I was collecting some leaves for my experiment, in this case hazel leaves. I noted that although the leaves were all green, some were far brighter and lighter green than others and this was because it represented fresh growth during the summer season. The leaves that had grown at the beginning of the season were dark green and tough, more fibrous, but the fresh growth was softer and more delicate. So, what we can see in the collected branch was that the fresh, new growth leaves had all been eaten and only a few of the older leaves had been bitten on – luckily, leaving the bite like patterns. This was a very useful indicator of selective feeding and by choice, the fresh leaves were favoured.

Here you can see a representation of the patterns left following a human eating a fresh leaf. Identical to the patterns seen in the collected sample.

7th. August, 2016

Report from Project Zulu:

> 'After recent events of the night of 1st/2nd. August, - we have now introduced Site 7 to our list. We are not in a position to talk about this at this stage, but just confirming that we have now broadened our research area. I wish I could say more on this, but for now it is best to just confirm that we have a new location with more strange activity.'

7/8th. September, 2016

Our first ever camp out over night:

> 'It's like they knew that we were turning up..... the exact moment we reached site 6 and put our kit down... we heard a sizable branch break through the trees and hit the ground with a loud thud. This was about 100 feet away from us..... we didn't see the branch itself, but saw some leaves falling out of the tree from where it came. Now... we cannot prove it was them.... But we are personally 99.999% sure it was from them. We went to investigate..... but could not find anything. So off to a good start. The night was fun.... Then things started to get interesting...... movement ahead of us running around... (no moon last night)... but it got the better of Ricky and he hit the lights. Nothing seen.. and it all went quiet after that. At about 1 am I was watching outside towards the Goldilocks zone and saw a faint glowing light... not bright flashes like we normally see. Ricky got some sleep but I wanted to stay awake as much as I could.... At around 3 am I heard bipedal footsteps behind our tent – about 20 feet away. I woke up Ricky, but the sounds just faded away. Nothing else to report other than a really nice dawn in the forest with an overpowering pine smell. Our first ever night camping in the forest and we feel it went well... more activity could have happened, but we were content with the confidence we gained and the action that we experienced. We hope to do it again now. One last thing.... When we were walking towards site 6 we got an overwhelming smell of honey or honeysuckle plants... we didn't check... but the smell was gone

in the morning??? Just an observation... don't know if it means anything at this stage as normally we only get the foul smelling odours produced by them....'

27th. December, 2016

'Just an update. We have taken everything off-line due to the nature of the study that we are now involved with. Those that don't buy into the paranormal side to Sasquatch would not understand the nature of everything that we are now dealing with. Fact. Our minds have been blown away by the latest and ongoing events that are happening. It is difficult for us to come to terms with this.... and for this reason, we are keeping quiet as we go much, much deeper down the rabbit hole. The whole BF community needs to open their minds up to a much bigger picture, the likes of which I would never have imagined when I first got into this type of research and why I am saying to everyone in this group - you need to think outside of the box as to who these Forest People are. As a result, we have given up trying to find Sasquatch.... that's all I can say.....

Chapter Seven

Paranormal Sasquatch

28th. February, 2017

One of the rare pictures of the Author and Ricky together at the night-time location.

One of the few last reports done on Project Zulu:

> Short update: Ricky and myself had our best night time investigation in the forest last Sunday (19th. Feb), proving the point that they are still active during the Winter months (which we already know to be the case, but this confirms it). Although we are keeping quiet about all the 'woo' related activity for now, which I will add, we also

experienced last week, I just want to confirm that Ricky witnessed a 9 to 10 foot tall heavily built sasquatch walk in front of us just 10 metres away. As it was completely dark, he could only make out the top half of the being against the skyline through the trees - which had broad shoulders and was walking slightly sideways on. As it walked past, it snapped some twigs on the forest floor - which we both heard clearly, alas, I didn't see it, but I knew it was there. Ricky wanted to hit the lights, but I told him not to and I started talking to the anglosasquatch in order to calm things down - as I can assure you, our adrenaline was pumping super fast by this point, but we held our nerve and thanked them for coming close and showing themselves in physical form. More happened this night leading towards actual contact. It was a very, very positive night and we will be back again soon.'

'It is rewarding to see that the whole 'Woo' side of things is now starting to gain some acceptance and people are becoming more open to this idea. For me, I can only stress that these forest beings sit outside of the normal status quo of the natural world as we know it and lie between our world and another. They have amazing abilities which are way off the scale to anything we would ever have imagined previously and we, as a team, are only just coming to terms with the gravity of this. I would like to say more, but right now I am still trying to analyse my own understanding of things and what has been happening to me personally and because I know that others would not take it seriously.'

I did a video of this encounter on Youtube – Sasquatch encounter 19th. Feb., 2017. The above report was put up on Project Zulu, but when things were starting to get into the paranormal world, I decided I needed to set up a new group that concentrated upon all of the weird stuff that we were encountering.

Below is an update from Ricky regarding this same encounter that was posted in the other group:

'It is with great excitement and astonishment that I write this little piece concerning the events of the night, Sunday 19th February.

The nights exercise began at around 6ish, with just myself and Paul, to be honest I forget the names Paul gives to each site (I'm sure he will fill in these blanks below) but we began with a little walk around site 4 which is situated directly from the car. Night was very swiftly setting in and the forest canopy was catching up with whatever daylight we had available to us but nonetheless with much anticipation from activity the nights before at Paul's Home we were both sensing a great deal of excitement and expectation for the night ahead, and boy we were certainly right to.

To begin, walking around Site 4 despite the lack of visibility at ground level we did immediately discover numerous fresh ground sticks as well as locations previously photographed by Paul showing signs of disturbance/tampering. Ground sticks varied in size, not so long or thick but each notable to follow a certain direction, you could kind of follow a route and predict the location the next one would pop up. It is also worth noting that we did pack lightly this night and carried nothing really but our torches and a couple of fold up chairs, so we did not take any photos of these ground sticks, however nothing that we haven't documented before, as common as the trees at times. Further on around 20 minutes into our walk around Site 4 I did discover a near circular assortment of branches diameter around 15/20 feet on the forest floor, this could easily be a natural occurrence by the way the long curved branches fell but still I felt worth noting in this report, but again with no camera we were not equipped to snap a picture of this for further scrutiny.

Upon leaving Site 4, we entered Site 6 and after around 15 minutes of trying to find our bearings we eventually found our usual base point. There must have been some moonlight in the sky as we could, after a short while, distinguish sky from canopy although of course we could not see our feet but silhouettes of tree trunks, branches and each other were clear enough from above ground level against the sky. Between talking to each other we tried an experiment of holding out our un-gloved hands and asking anything around to touch them but neither myself nor Paul got anything back from this experiment (worth a try). The first contact we did get this night was instead during a moment of break, talking to each other about something (can't remember what, but unrelated to Sasquatch) all of a sudden I felt a tiny grab on my elbow, nothing too pressured but firm enough to flinch and fall silent thinking it was actually Paul silently communicating to me that he had seen or heard something, but no it wasn't, instead asking me what's up. Weird, but was enough to put me on edge for a bit. Perhaps it could of been a trapped nerve, at least that's what I told myself, but it was enough to warrant calling out to ask if there was anything out there wishing to make contact with us to do it again. Nothing once more, so was in two minds of writing the event off, however shortly after calling out chatting once more about (unrelated matters) something happened again, this time to Paul, a tap on the head while stood up. It seemed to be from this point that calling out and other experiments were void producing any concrete results and the entity seemed to only want to interact with us when we were in fact relaxed and unexpecting of it. Nonetheless we decided to start speaking out once again, sympathising with the forest people, and trying to build more common ground and understanding. Talking for around 10 minutes Paul opened up to our friends and reassured them of our presence in the woods as well as other topics for instance our mutual distaste of the deforestation present and other wider human environmental impacts on the planet, between while requesting a sign from our friends that they were still out there listening. Nothing again happened when we asked instead however for the first and only time during the night an owl screech sounded from the distance seconds after we ceased speaking, gliding through site 6 and disappearing.

Now to the business end and the most significant event and reason I break silence and write this. Half an hour or so later, after the owl screech, whilst talking to Paul all of a sudden and hand on heart I caught the moment we had all been waiting for, an actual visual sighting in the shadows of what must have been a 9-10ft Sasquatch about 10 meters away from us. Due to visibility I saw nothing from the waist down however I clearly witnessed the upper silhouette against the moonshine (head, shoulders and arm) walking side on from left to right between two nearby trees. Safe to say, as Paul as my witness I was pretty scared and grabbed Paul and went straight for my torch, but I don't know how, resisted the temptation to turn it on. Seconds later a twig snap from the area was heard and we both then sat in complete silence The event happened so fast and unexpected that it is difficult to say for certain but the creature seemed to be gliding not walking, although I did see the shoulder part of an outstretched striding arm. Now sceptically it was dark and getting late, and although now and again eyes play tricks on you in these conditions, building an actual life size image of a Sasquatch had not ever happened to me before and accompanied with the twig snaps that followed in the same area could only reaffirm that this experience actually happened, and I firmly believe this was not a trick of the mind. I have included a quick sketch below of what I recall; not the best artist I'm afraid but it's something. I honestly can't believe how close they came to us in physical form this night but now it is clear these guys are the masters of stealth and can be but inches away from us but remain obscure and concealed to the unbelievers and average Joe public.

Ricky's artistic representation of the figure he saw passing by us on the 19th. Feb, 2017.

25th. May, 2017

'Report from last night 25th. May, 2017. Another good night and with some new forest records being made. We believe we have recovered a juvenile print (video below). I will do a video later of the cleaned-up version. Alas, the toes have not been preserved very well in the plaster print. I noticed this when I lifted the print that the plaster had not fallen into the gaps for all the toes. Oh well... Hopefully there will still be some useful information we can gain from the print, but feel we have recorded a footprint here.

We also found the tallest ground stick ever - we estimated it at being in the region of 35 - 40 feet. It was shoved into the ground and in association with the tree arch and no stump - there was no other reason otherwise. I noticed in the video (see below) that there was also another branch on the ground that was first used to hold down the tree arch... but didn't notice it as I was blown away by the size of the huge ground stick. This was a record breaker and proves the point how strong these forest people are.

As for the night investigation... it got off to a slow start until the woods got totally enveloped by darkness (new moon so very dark, but with brilliant starry night). The first thing that we heard was huge and I mean HUGE stomping on the ground about 50 to 100 feet from the left of us. It was like a human bashing their feet on the ground but times by 10. Next there was a rush through the undergrowth - then silent - then from our right there was movement and some grunting sounds. Yes.. our hearts were pounding at this point - then out of no where a huge plane (army) came and flew straight over our heads real low - just above the tree tops. Then a second one - but a little further away. How does this happen? Just when things started to get fun! After that we heard a muntjac deer barking in the distance to our right. OK.. it happened again... military craft blowing out the action just at the point things were getting interesting. I will put it down to another coincidence, but if this happens again like this (and I mean just at the point we get action), then I will come to different conclusions. Those planes killed off the action for the rest of the night. We did hear twig snaps - but at distances over 100 feet away and nothing really happened.

After we collected the print - we were uncertain if our marker stick had been knocked over by them in anger at us taking the cast.... we are

Print recovered from a small muddy area located within site 6.

unsure. (video to follow). But we were sure that the stick (with the plastic bag on the tip) had been pushed over. That was left there to help us find the print in the dark... but it was lying on the ground! Following the collection of the print - for the first time we felt like we were being watched as we left the forest - an uneasy feeling this time - like we were being escorted out of the forest. We think they were not happy about us recovering the print. That was a first in a long time of that uneasy feeling, but I am sure we will be back and everything will be OK again.

I will add that maybe there has been a change recently there.... as the first thing that I noticed when we arrived was that some of the marker sticks had been knocked over. So maybe they were not happy for some reason before we had arrived??? I don't know. But it was an interesting night with some great new finds.

24th. June, 2017

'Back from the forest just now and given this was a daytime visit.... I have to say that this was awesome. I feel they were watching me from a distance as I made a very long video about revealing a few things about what has been happening over the last 12 months. Within a minute of finishing the video I heard a wood-knock... we (Gem and myself) stopped and listened and then there was another loud crystal clear wood-knock - then I got the camera out.... and managed to record the last wood-knock. Very clear and close by.... I have only ever heard a couple of single day time wood-knocks... so this was a first and it was right after making that video (which I hope to upload later).'

16th. September, 2016

Creation of the new group – 'Project Zulu – the X Files'.

'I have created this group for the team, with a potential for a selective number of other people in the future to be added - agreed to within the group. Given the experiences that the team have had following the whole 'WOO' type of events, I don't think we should share these with the wider audiences and keep it to ourselves as we progress further into the unknown. I thought best to create a secret group to document all of this.'

16th. October, 2016

First time in the home experience:

> UPDATE FROM LAST NIGHT: Although I could not get out to the woods last night... it certainly didn't appear to go without some sort of action. At 2.20 am this morning, whilst Gem and myself were asleep - the TV in our bedroom came on by itself. No, no one was near the remote and no, this has never ever happened before. Yes, I had been meditating the evening before. The question is: do tv's come on by themselves at random times? or can we say that this is yet another example of them interacting with us? We will just have to wait and see if this should ever happen again... if so... then yes, I would have to say that they came to visit me again in my own home! Awesome!

For the record, any references to the new Site 7, then this is regarding our own home.

3rd. December, 2016

'Looking at these things (Orbs) historically - in the UK they were once called Will-o-the-wisp. The term "will-o'-the-wisp" comes from "wisp", a bundle of sticks or paper sometimes used as a torch, and the name "Will": thus, "Will-of-the-torch". They were believed to represent spirits. One version, from Shropshire, recounted by K. M. Briggs in her book A Dictionary of Fairies, refers to Will the Smith. Will is a wicked blacksmith who is given a second chance by Saint Peter at the gates to Heaven, but leads such a bad life that he ends up being doomed to wander the Earth. The Devil provides him with a single burning coal with which to warm himself, which he then uses to lure foolish travellers into the marshes. Image (not included): The Will o' the Wisp and the Snake by Hermann Hendrich (1854-1931). Isn't history related to this subject fascinating!'

19th. December, 2016

'I never really had any interest in Dallas & Wayne Bigfoot Hunters due to the way they were portrayed in Shooting Bigfoot (which was fun, but poor media). Alas, Dallas has recently passed over and I have been watching some nice tributes to him and that he was the real deal. So, I have just been watching some of their episodes 'The Bigfoot Hunters', and it became apparent that they were closer than most here and I learnt a few new ideas to use myself. Now... one of the first things that I picked up upon is the fact that they said that the sasquatch can appear and disappear in a bright flash of light. Now... the Patterson Gimlin film is the only film footage that appears to show this (that I am aware of)... apart from my recording. Are we sitting upon a powerful piece of evidence here that can help move the sasquatch study more forward into accepting the paranormal side of things? What is the best way forward here? I was told to sit on it... but is the time right yet to make it known? Ideas guys? We are already witnessing the orbs and flashes of lights in the forest on a regular basis.... together with the reports from John - so, we already know ourselves that they are connected. But the mad BF world out there are so slow catching up on this... and very hostile too. Merry Christmas!

Anyway, here is the episode where Dallas swears that he saw a sasquatch disappear in a flash of light. He is not the only person to report this... but just saying.... '

YOUTUBE.COM
Dallas & Wayne: The Bigfoot Hunters - Episode 1 - Segment 2/2

Worth a watch if still available, where Dallas
confirms that he saw a Sasquatch disappear in a bright flash of light.

25th. December, 2016

'2:25 am this morning I was woken up again to the TV being switched on. This I now understand is their calling card.... I got the remote, switched it off, then checked the time - again, exactly at a similar time, if not exact, time as before. Anyway, this time I knew it was them and I got a telepathic message - it's like they know what is going on in my life right now. The message came through loud and clear - 'the road may be tough - but the destination is real'. Following that my emotions went all nice and I drifted off to sleep happy. For those that don't know, the biggest issue I have right now is (deleted out). Anyway, I gained so much comfort from this... and it has made my Christmas already..... I'm off to the woods shortly to take Toby for a walk and will be saying a massive thank you. This whole Sasquatch thing really is taking a on a different level, but nobody out there would ever take it seriously.... you have to experience it personally to understand. Have a good day everyone!'

18th. February, 2017

'Last night, guess what time? The TV got turned on again. This was the first time since Xmas morning. The reason why? well, I have been making some progress in mind-speak with these people and last night, after chatting with Ricky on FB, I asked them if they would come and visit that night. Well, well, well... what happened? OK, just for the record, if you have not already guessed... the TV came on at 2.23 am - now that is about the same time as every time before (well, the last three times now) around 2.25 (ish) am. It made me smile... and I said my thanks to them and then drifted off into a deep sleep with a nice dream being at a theme park having fun. Was this a prediction for me? I sincerely hope so... I can now safely reject this having anything to do with a malfunction on the TV - this is beyond doubt their contact with me which is in the realms of quantum physics. How awesome is this??? Gem still can't get her head around what is going on and chooses to reject it... it's easier for her to deal with that way.

After the TV turning on by itself last night, this being the 4th time now at 2.25 am (ish) and having pressed Gem today... she finally admitted that yes.... This all relates to the BF connection. She has been holding back, and she doesn't like the unknown, but secretly she understood this to have been the case for a while now... but didn't want to admit to it. She has accepted it as being part of my life now and that more strangeness may well happen in the future, but she supports me with this.'

21st. February, 2017

This is a follow-up additions to the report above regarding the encounter on the 19th. February:

> 'Point to add.... Not had much chance to talk things through fully with Ricky yet, but it seems that since the encounter, both Ricky and myself are feeling really relaxed (more so than normal etc.), I myself are feeling happy and content, even though I have a few ongoing worries on my plate... I just feel that something changed after our visit. Anything to add to that one Ricky? Yesterday I felt quite happy and full of fun....
>
> Ricky: Absolutely agree, I've never been happy really in my job however since last week I have gone to work relaxed and mellow with an attitude of 'oh well, who gives a ****' when I start stressing! Been cycling into work all week too and refusing the convenience of a lift I've been getting from outside my house for years! Somethings changed in the air, life simply doesn't feel so serious anymore'

26th. February, 2017

'Lets face it... whether you start talking about multiple dimensions, cloaking, portals, zapping, switching TV's on, turning cameras off... you are talking about changes in energies. So, it has got me thinking as there is a lot of talk about stick structures right now etc. OK, at Site 6 & 4 we see a lot of ground sticks in lines... are these some type of energy lines? So.. next time I go there I will be trying to use some dowsing rods and walk the area and see what happens when concentrating upon changes in energy patterns. Maybe there is something there that we can pick up upon... good idea?'

I ended up trying dowsing rods and had some success with this... not just with myself, but others who didn't even understand the significance of what a ground stick was. It was not conclusive, but certainly there was something there that required more analysis.

1st. March, 2017

'OK, in order to document everything that is out of the ordinary of late... I want to add that just recently, and I mean, just in the last week, when taking Toby for his morning walks I come across Robins perched on branches just 6 feet or less from me singing without a care in the world that I am standing right next to them watching. This has happened three times now during this period. Once would be rare event to have that, but three times in the last week is incredibly strange. Either something is happening to me, in that I no longer appear to be a threat (like with the deer following the encounter)... or something else. I know Robins can be tame, but this must have been three different Robins as at different locations. It's a strange one... but one that I thought I would highlight as it appeared way out of the ordinary to me. Oh, that reminds me... the day after the encounter, I was walking Toby down the fields behind us and I counted 18 (maybe 19) magpies all flying together as we had disturbed them on the ground. 1 for sorrow, 2 for joy etc... so what is 18? Seen plenty of magpies together, but that was a new record, plus it was the day after the encounter - jinx? Strange, but true. I don't know what is happening right now, but I do feel something has changed within me. Apologies for this, but weird things like this I feel need to be documented.'

There may be some connection here to what I will mention in Part 2.

22nd. March, 2017

'OK, it was a sad day for this country today. We were all in shock by the scenes that we witnessed on TV. So what has this got to do with this thread here? Well... I swear on my life this actually happened. At about 2.30 pm today I was getting ready to take Gem out for afternoon tea to celebrate her birthday. I can't be sure of the exact time, as it never occurred to me that anything was wrong at that point, but only now do I realise the significance. We left the house at 2:50 pm. At some point between 2:30 and 2:45 pm the tv in the bedroom switched itself off. I didn't see it happen... but the TV has an automatic power down facility after a number of hours viewing without pressing channels/ volume buttons etc... so, I automatically assumed this had happened. To my amazement - 5 seconds later the TV switched itself back on again. I looked up to see how that could have been done and could see the remote on the bedside table well clear of anything - so, it wasn't Gem or the remote responsible for switching the TV on. This is the FIRST ever time this has ever happened - where the TV switches off and then on again shortly afterwards. For me, at the time I just thought that was a little weird but never thought anything about it as we were rushing around to get ready to go out. I never thought about it until 10 mins ago - due to all the news coverage my mind was elsewhere.... but then I recalled what had happened earlier on. Just checking what time the incident happened in London I was shocked to see it was 2:40 ish. Faced with that, I have no other conclusions to come to other than it was these forest people notifying me of something that was just happening or about to happen. Just like with the Trump elections at 2:15 am - this time it was related to a main news event here. This is just confirmation that these forest people are way, way smarter than we give them credit for and they are certainly communicating with me here at home in a very unusual, but predictable way. As I say, I swear on my life this actually happened. I'm blown away yet again.'

14th. April, 2017

'The last 2 nights I have had a visitor.. Last night can you guess what time he turned up? Yes, again, 2.25 am switching the TV on (BBC 1 - all about the prospect of life on one of Saturn's moons). Then it was a case of mind-speak for quite a while... all encouraging things came through (some not what I would like, but none the less positive for the future (personal life related). Most things I asked got turned upside down on it's head... so I know it wasn't me making it up in my head. Now... the previous night I had one visit me during the night (no idea what time it was as I was half asleep) but he touched my neck and twice to my feet that were sticking outside of the covers. My mind-speak then was strong and short (even though I was half asleep). He said he was quite happy for me to pursue whatever angle I wanted with regards to them... that there was now a bond of trust between us'

16th. April, 2016

'Talking to Ricky tonight in the forest prompted me to confirm something after our discussion. This week, I got the TV switched on again at 2.25 am. I know its not a fault with the TV as it has never happened before and happens at key moments (Tumps election, possible life on Saturn's moon, Terror attack on London and Xmas morning message just for me). Key moments yes in recent months history? OK... why basically at 2.25 am? Every time bar the terror attack? OK... The TV I have has no time program, yet it comes on at 2.25 am. The clocks changed last week to British Summer Time... and yet the TV got switched on still at 2.25 am this week! Gem will confirm this. So.... how is TIME based upon in their world? In accordance to our understanding of time.... or.....???? Is 2:25 special for some reason? Answers on a postcard please......'

18th. May, 2017

'If mainstream science is openly talking about multiverse and parallel universes then it is not too much of a leap of faith for us to be openly talking about what we are dealing with here. It is fair to say that what we have encountered is, in essence, seeing the physical effects to the physics behind this.'

23rd. May, 2017

'Sick and sad news from last night. Again, a coincidence? A few mins before the news came through..... having not watched the news all day... maybe even for a few days in fact. No, the TV didn't switch itself on by itself this time.... but having played football last night I was relaxing, just using my laptop.... then, for some unknown reason I wanted to switch the news on... not just the TV - I wanted to watch the news channel. At that exact moment there was nothing on the news - but within minutes - possibly just 2 mins... they interrupted the news to say that they are just getting reports of an 'incident' in Manchester. That was just mins. after I had switched the TV on.... I can't explain it... but I felt I wanted to switch the TV on and watch the news - I don't know where that idea came from... but the timing and urge was significant. Again, was this just a coincidence - well, for me.... no.... and that is why I feel it appropriate to document it here. Just like the March 22nd. attack on Parliament - significant news items are being highlighted in some way to me. That is how I feel and yet I can offer no proof of this.'

28th. May, 2017

'Had a visit last night at.... You guessed it.... 2:22/2:23. It was 2:23 when I looked at the time on my phone, but could easily have been 2:22 as it took me a short while to find it. Gem was asleep.... So I didn't bother to wake her up. What was on TV? Some replays of wrestling matches from when I was a kid - big daddy and giant haystacks.. weird...'

2nd. June, 2017

'Guess what time I had a visit last night? Yes...exactly the same time again. 2:22/3. They are so punctual it's just so dam amazing!!!! Had some mind-speak too... I'm not so sure what the message on the TV meant??? Will see if that has any meaning that they have gone away for a short time or not???? Blown away.'

14th. June, 2017

'London again having to face a new tragic situation, this time a Tower Block Fire. Totally horrific scenes. I cannot imagine what it must have been like for the people caught in that fire. My reaction? Well.... it is rare for me to remember any dreams, but the night before last night I had a weird, weird dream which didn't make any sense to me, but now I cannot dismiss it as pure coincidence. It didn't occur to me to write about it at the time, as it was just odd, nothing more than that to me. My dream was about me finding some bigfoot footprints around the base of a church (the church from my home village, although parts that I didn't understand). It didn't make any sense at all to me... I found all these footprints at the base and then I wanted to go find some plaster so I could take some casts. When I got back with the plaster... the whole of the church (and it's Tower) was on fire in a huge blaze. I was in shock seeing the scene... and that is the point I woke up. I didn't think it worth while for me to mention this at the time - despite this being a weird dream and bigfoot related - it was just a strange one for me. And then last night.... nothing switched on... but for some reason which I cannot explain - I could not get to sleep. I was awake from around 1 am through to 4 am and as much as I tried... I could not sleep. This was completely unusual... but I just put it down to my over active mind. Only switching the TV on this morning did I see the news. I have to ask if these dreams carry some hidden meaning... again, I cannot discount this as a pure coincidence. The timing again was too specific, although not quite relative to the tower block - but the Church Tower may have been a symbol to me that I would relate to afterwards??? I don't know... but hand on heart, I had that dream the other night. If I get any more weird dreams related to Sasquatch (and I add, I don't normally remember dreams) then I will do my best to record this and see if anything happens as a result. Another sad day for London.'

26th. July, 2017

'Just to let you know that I am currently communicating with someone who works in the Forestry Industry in this country and he has been showing me pictures of classic Sasquatch wood structures. He has confirmed that some have appeared overnight in private woodland etc. He also said of an account where they had loaded up a trailer of tree trunks ready to be transported away the next day from a private wood. The following day when they turned up, half of the tree trunks had been removed from the trailer and used to block the entrance for the collecting lorry. No tracks for any vehicles to have done that - so, he was left very confused, considering the sizes of the logs in question. This is a useful contact and gives a few more ideas about the strange behaviours that our forest friends get up to when their woodlands get cut. I will keep the group posted if anything extra is learnt.'

27th. July, 2017

'Well, yesterday I mentioned that they have not visited me in a while. I guess they must have heard that as, as you can imagine the time, they visited me last night. Again, absolutely crazy yet true, the TV switched itself on at exactly 2.22am. Gem slept through it, but I was woken up despite being in a nice deep sleep. I didn't get anything from the visit, it was just like they wanted to tell me that they were still around. Now, was that because I wrote about it last night? Well, in my own mind that is exactly what the case was. I think I need to go back to the forest again real soon to say hello. Well, that is my report from last night. Again 2.22 am. I'll give it to them, they are very precise and punctual.'

29th. July, 2017

'What are the chances of a film coming out this year called 2:22? Well, that is the case and I had to watch it in case there was anything coincidentally included in it. Sadly I didn't pick out much, but sense there was something to learn. Worth watching, even if it doesn't make any sense as it talks about patterns, connections, coincidences, convergence and also a cosmic connection (due to a dying star). I think it will bug me for a while, but was it just a coincidence there being a film by that title this year? Answers in a postcard please..'

16th. August, 2017

'OK... something is happening. This is a first. The last 2 nights I have been visited with the TV being switched on both nights at - care to guess? - 2.22 am exactly. (Tv programmes not specific - London athletics 1st night and just a financial report last night). I had mindspeak on the first night, mostly personal stuff and last night it was just weird feelings then I fell asleep. Will wait and see if anything happens again tonight.... but feel this is building into something bigger. Will keep everyone posted. I will add that the mindspeak is getting stronger and more often - but still very much a one way conversation with me asking questions and getting very limited responses.

Hope to be getting out to the forest on Friday for a night time visit - (still up for this Ricky?). Fingers crossed we get contact....'

19th. August, 2017

'Last night Ricky and I did a night time investigation - alas, the weather report wasn't correct and we dipped out early as it started to rain. The wind and the rain together meant we had no abilities to sense in the darkness. Before it got dark we went on a search around site 6. At one point we decided to whistle and do a few hand claps. Amazingly we got three wood-knocks in response. A single one then a pause of a minute or two and then two clear wood knocks within 20 seconds. They were still a little away from us, but clearly in the forest. We then returned to our normal spot and waited and chatted. Alas, the rain and wind stopped play, but before than happened we talked about the significance of the

2:22 timing I get at night. OK... how we came to this was bizarre... but we were thinking outside of the conventional box... so... take this how you will. Bi - is also reference to 2. 2 bi 2? or 2 by 2? Was this in reference to a biblical reference for saving the creatures of earth like Noah did???? I don't know... but 2 by 2 made us stop in our tracks when we said this. Where the thought came from? Pass.... but 2:22 and 2 by 2 could have a symbolic connection??? Thoughts?'

16th. September, 2017

'Back to our location last night. Nothing exceptional to report, but they were with us, but chose not to have much interaction with us. We both saw a number of orbs - but this time, they looked different - they were closer to us, but they were much smaller orbs (like fairy lights is the best way to describe them). I saw 4 of them last night - the closest one was just 30 feet from us in the tree canopy just above us. Awesome - alas, nothing more. The orbs lasted just a few seconds. Towards the end I was certain I was hearing talking and whispering just ahead of us - it didn't help as we had rain droplets falling from the trees.... but I am certain I could hear them talking around a hundred feet ahead of us, alas it didn't amount to much more than that - but they were close to us last night for sure.'

18th. September, 2017

'Got a visit again last night - yes, at 2.22 am again (pic. taken at 2.23 as I did that as an after thought). Powerful mind-speak which I am still trying to digest this morning. I am feeling a little weird this morning, which may or may not relate to what I was told - as they said that they would be giving me a gift(???). I am in awe.... they are true kind spirited people. I am uncertain if there will be another terrorist attack (these visits can be a precursor) soon or if this visit had anything to do with the visit we had to the forest on Friday night. But I know that the contact I have now is at it's strongest point achieved so far to date. There is no going back now. In answer to John's idea of there being two beings - I guess I am in agreement on this now, as I was told that my communication last night was with the 'star people'. The connection between the star people and the forest people is a close relationship. The rest I will have to sit on as I need some time to think about things.'

18th. September, 2017 – image captured of the TV being switched on at 2:22 am. Time of picture: 2:23 am with a forest background.

25th. September, 2017

OK, just to report I had yet another visit over night at 2.22 am this morning - again. I also had quite a bit of mind-speak too. Why last night? This has nothing to do with any terrorist activity. Yet, right now I will not say why it happened, but for me, this confirms everything that what I am dealing with is indeed special and not because of any issues with my TV. I hope to explain why it happened last night soon - but right now I need to keep this in my own mind. But please understand that this really is happening and yes, this is leading up to actual contact soon. These are crazy times. Still no idea why 2:22 - but they have repeatedly come at this EXACT time again and again. Amazing??? I just wonder if they do this with other people at this exact time??????

Then later:

> 'May I welcome Gordon to the group. Gordon has had some awesome experiences at SOIA a few weeks back and the activity

has followed him home. This is ground-breaking developments now and I hope I can gain some more meaningful contact with our wonderful forest friends with Gordon's help. Welcome Gordon!'

27th. October, 2017

'Having been away a while on holiday, I came back to find that on my first day back that our forest friends had left me a sign whilst I was at work. It is quite funny, but they pull the sun visor down in the car (which is naturally locked all day). This has happened quite a few times now. Then last night I had a visit again at 2.22 am together with some meaningful mind speak. It was incredible. The programme on TV? Jackpot! Anyway, that's what I am hoping for very shortly - i.e. hitting the jackpot with actual contact with our forest friends. Fingers crossed.

Whilst I was away, Ricky stayed at my place (8 days) and he was waiting to see if the TV got switched on etc. alas, it didn't during his say however, on his first night Toby (our Labrador) started to bark at nothing - and the time? 2.25 am. That was the only time he did this as he has never done that before either when I am at home or following that first night. Weird? Yes.... but I think I know the reason for this..... Anyway, I'm looking forward to next few weeks as I am sure that there will be some awesome updates...'

6th. November, 2017

Just a quick entry... I was visited again on the Wednesday night/Thursday morning just gone (1/2 November) at again, 2.22 am. The contact is getting more regular now, which is great. Yes, I got some mind-speak too - and this relates to my meeting up with Gordon on the Sunday night (29th) - which was awesome. OK, i will add that I only saw one orb near by while we were there, but we were in company with two groups of forest friends - those from Gordon's area and the tall one from our site. Gordon was trying to teach me how to see them, alas, i was struggling with this as my eyesight isn't great.... but I stood just inches in front of a 7 and a half foot Clan member from Gordon's area and talked with him for a while... no sense of fear, in fact, quite the opposite. I missed the yellow eye glow when Gordon had asked him to reveal himself to me... but I could sense he was there, even if I could not see him. I will ask Gordon if he would like to add his report from the night here. The night was, even if the activity was limited, it was still INCREDIBLE with a capital I! The mind-speak I got on the 1/2 November backed up a number of things for me regarding that night.... so looking forward to the next encounter!'

6th. November, 2017

'Last night was interesting. With a few things kicking off in certain areas right now, it came as no great surprise that i was visited last night. The interesting thing about this visit was that I got min-speak first at around 2.20 am and as a result I had opened my eyes just at the point the TV switched itself on! At 2.22 am... The TV went a dull green for a couple of seconds and then the picture came on. I continued to get mind-speak (especially in relation to current events!). This was all welcome and helped to settle my mind and made me smile. The interesting thing was following this. I turned the TV off. Then waited a few minutes and thought... having seen the tv switch itself on, I wonder if it switched on in the same way as I had just seen. This was the funny thing... the TV wasn't right, it was going blank, the picture came on and turned itself off....and then coming on. Now that wasn't right. It has never acted in that way ever before, so I turned it off again and tried it again. The same happened again! Totally bizarre. At that point I knew something strange was happening with the TV. This morning when I woke up for work, I switched the TV on and the picture just came up and switched off normally. This is very different to the way the tv responded at 2.22 am. So... for me, it is clear the TV behaves very differently at 2.22 am to when it does normally. I have been asked to swap the TV's over to be double sure... but I don't feel the need to do this as this episode just confirmed things for me. The fact I get mind-speak at the same time is confirmation in itself... Naturally gem slept through the event as normal.'

9th. December, 2017

Yes, it has been quiet for the last few weeks or so... but for good reason I have not mentioned anything about my visit to Gordon's area - which was awesome. That took place on November 18th. I will ask Gordon if I can copy his report from his group to show here too. Anyway, since then - 3 weeks - I have had no contact, no night-time visits, no TV's switching on at 2.22 am, no mind-speak - nothing. Quite a sad situation for me and I thought I had done something wrong to upset them and as a result they have kept away...... So it is with great pleasure to report that following my visit to site 6 tonight with Ricky, we had the best night for ACTIVITY there EVER. We arrived, wanting to remain in TOTAL darkness - and considering the long walk up to site 6 - we did well sliding around in the muddy track to get to a point where we thought we best put the torch on to get into site 6. This is where we laughed. Now, anyone who has gone to site 6 will realise that it is a fair trip there and not so easy to get there.... so imagine, to our surprise, that the moment we switched the torch on - we were at the EXACT spot that you enter site 6 along the path we use!!!!!! That was an impossible task in the dark and yet somehow... that is the point we stopped at and switched the torch on - unbelievable! So... off to an incredible start. It was cold tonight, real cold - minus 1 - so we were not expecting much in the way of activity as most happens in the summer - but we wanted to show willing. It was quiet to start with whilst we chatted for a short while and then I stood up and started to talk out to them. Then Ricky stood up and joined in. At this point we saw the odd orb in the distance. I then tried to explain to Ricky how to see the shimmering outlines of the forest people when they are close by... for which Ricky was starting to make out the odd moving image in near invisible form - he wasn't sure, but feel he was actually seeing them and close by too! I said we will get Gordon back so that we can confirm this with him when the clan are directly in front of us. Ricky also got a tap on the back of his shoulder. So... it was starting to turn into a good night. I then said lets play some music for them and at this point - things got more lively. Our moods became more upbeat and likewise for everything around us. Here is the weird thing - Ricky's phone was reduced to 1% battery life (in the freezing cold) and yet it continued to play 6 songs on youtube! Ricky did not understand how that was possible as his phone always dies straight away at that point - yet somehow his phone continued for over 15 mins! Where did that power come from?? Anyway... this is when things became more interesting - we started seeing orbs in all different directions - all surrounding us. It certainly wasn't our eyes as both Ricky and me were both seeing the same ones. Some bright - some flying between the trees - others like small fairy lights. One went off just 8 feet in front of me close to the forest floor. I heard whispering at one point not far away and we were hearing twig snaps ahead of us. Red, white and blue lights/orbs. At one point it felt like firework night as we were each seeing them in different directions every 30 seconds or so. I would not like to guess how many we saw - but well over 100 I would say. One repeated red and white lights confused us thinking it was a car in the forest at a distance (but how?) as they were bright and sometimes lit up the trees - yet after the night we went to investigate if it could have been a car and we came to the conclusion - this could not have been so. In that case then something bizarre must have been going on at a distance from us which if it happens again next time then we will go investigate. We called out and thanked them for the show - and I understand I got some mind speak for the first time in weeks. I asked if they will visit me again and I was told to wait and see. I asked if it was OK for Gordon to return and it was a definite yes. So, we stayed about 2 hours and about an hour and a half we were treated to an incredible display. Progress has certainly been made tonight. Most orbs were around 100 feet away... some greater distance and yet some were real close up - within feet of us. Alas, the cold got to us as we were prepared, yet not prepared enough. We left in high spirits and said we will be back again before Christmas. Anyway... I will see if I get night-time visits again - but we feel that everything was very, very positive tonight and was a night like no other we have ever had there. All feeling happy here tonight!'

29th. December, 2017

'OK, I am now opening up about various things that have been happening over the last year. I have even opened up a bit about things on Project Zulu, so I feel it is only fair now to open up regarding what else I have been holding back upon. As I have already mentioned before - I do get mind speak, and on a regular basis too. Hard as it is to imagine, it is happening. What I have not said before is the name of the contact that I have - he is the clan leader from the forest area we go to and his name is Zachariah or just Zach for short. Generally speaking, the communications have been a one-way road - with me being told things on a need to know basis. However, that has all changed since my dealings and involvement with Gordon. As I have already mentioned, night-time contacts have all disappeared in the last couple of months - no TV's switched on a 2.22 am. Well, there is a reason for this, and this will be explained in Gordon's reports (which I have copied over).

Now, generally speaking, I do have an open line with Zach and can communicate with him on a daily basis. Yes, this does sound far fetched, and I understand that, but I am just reporting back upon my own personal experiences. Anyway… I will record Gordon's reports in the manner that they took place. We have only arranged two site visits and both have been rewarding successes.'

Chapter Eight

Another Year, New Experiences

Humans can only see 1% of the electromagnetic light spectrum. 1% only. This means that 99% of the light spectrum goes unnoticed by our eyes. So, there is a lot going on around us and we wouldn't have a clue about it.

And do you know how we are reminded of this?

A picture taken of a beautiful rainbow whilst on holiday in Devon, 2022.

The white light given off by the sun, when hitting a rain cloud can generate a rainbow under certain circumstances. What do the colours of a rainbow represent? Well, these colours represent the limitations of our own visual abilities. The violet, seen at the base of the rainbow is the visual limits for our ability to see ultraviolet rays, and on the outer side, you can see red... again, our limitations of seeing infra-red. As an experiment, if you have a black and white camera fitted without a UV filter, then take a picture of a rainbow, one in colour with your normal camera and the other with the black and white. You will probably notice an extra band of UV light underneath the one that can be visually seen with your own eyes. This proves the point of our own visual limitations.

2nd. January, 2018

'Wow... ok, this one goes out there with all the other weird stuff... but I am now being open about my own experiences dealing with my contact - Zach. I had a super long chat with him last night and to cut things short these were the main points - I talked about the universe and was told there is no such thing as 'matter'. Everything and I mean everything is composed of energy and depending upon the frequencies this can either vibrate at a higher or lower level. The more 'solid' the energy, the lower the vibration level. We are a mix of mostly low vibrations and also some high. The lower is our biological 'container' which restricts us in many ways and our soul or inner being is on a much higher frequency. This is the difference between our kinds. A lot more personal stuff was talked about... all good by the way, so happy with that. There was a lot more... alas, by the morning I had forgotten most of it...

As hard as this might appear... I am experiencing this mind speak..... and it's incredible!

Just remembered one more thing... I asked if I could see Zach - he said 'why was it so important? Is seeing - believing?' I answered - yes. He then said - then is hearing - believing? And is feeling - believing?' if so? why do you need to see then? I just answered -you are correct. I have five senses and I have smelt, heard, touched and felt your presence - but one day I would also like to see you so I know who you look like and he then said - 'you will'.'

6th. January, 2018

'Report from tonight (5th. Jan, 2018). I have to say, it is much nicer doing the research in the summer time - as yet again, it was a cold one - at around freezing point tonight, but this time it wasn't the cold that got us, as we were better prepared this time around. So... tonight we did things very differently..... and it gave results - although they lasted for around an hour then everything went quiet on us. So, now that we are looking at 'energy' in a new way related to these forest beings - I said we would go for a walk around a prehistoric Stone Circle before we would travel to the forest... so, that's what we did. Upon arrival at the forest, we sat down with two beers and wished our forest friends a Happy New Year! Soon after this I could see a shimmering image just to the left of me - at around 9 feet tall. I got up and then felt the top of my head touched. I went over to the 9 footer and held out my hand, but I got no electric handshake this time, but did see a very brief yellow eye glow. Ricky came and did the same and he felt something like the sensation of a money spider crawling over his hand. Then out of the corner of his eye he saw a smoky grey image appear behind my back (as I was bending over to put the torch down in my seat). This was right between the two of us and we were only 10 feet apart. I had my back to him, so I didn't see him. This said, I believe it was before this, or just afterwards - I can't quite remember - but Ricky and I was standing together in front of the 9 footer and then we both saw a 4 foot shadow run past just 15 feet from our left. Like the orbs - if one of us saw it, you could imagine it was your eyes playing tricks with you - but this time, both Ricky and me both saw it run past. An amazing feeling when you know that they are that close and revealing themselves. During the night we saw around 25 orbs in total - a few also that both Ricky and me saw at the same time. We both saw some other shadows/figures at various stages and also Ricky was also telling me about how he is getting his first mind speak - and wasn't sure what was happening to him - as he was hearing these voices inside of his head calling 'Ricky'. He was starting to think that he was loosing it, until I explained that they were starting to mind speak with him. Next time it happens, he will sit down and focus on them and speak back! We played some music and put up some coloured fairy lights for them - which I feel went down well... then quite suddenly - everything just went quiet - no orbs, no shadows, no shimmering - no nothing really. So we chatted for another half an hour then went and made a video on some fresh ground sticks and a glyph we found. Overall, yet again, a very, very successful night! Did the energy from the Stones have an effect? Don't know - but again, they were close, real close to us and revealed themselves to us - which was amazing.'

Just one other point to add. Considering the air was still, absolutely still – at one point we got a massive blast of lovely pine smell that hit us for a short while – around 30 seconds or so. We both smelt it and both knew that was not normal. So, it appears as well as the foul smells they can produce, they can also produce the lovely smells of the forest that we love. For the record, it was an overwhelming intense smell of pine.... And no way was it just normal etc. Later, when sitting down we both picked up the lovely smell of – well, the best way to describe it is you know when you blow out a candle (a pine fragrance candle) and you can smell the burnt wax/pine smell? Well that's what we got and it was by no mistake as that is what both Ricky and me stated. Again, a lovely smell that came from no where on the still air. How do we prove it was them? We can't, but that was our observation and thoughts. How do they do it? Again, no answers for that.'

24th. February, 2018

'OK, where does one start? For some people it is difficult to relate to what we have to say here.... as it defies our understanding of just about everything. We understand this, and to a certain degree we can relate to that as it is off the scales to our understanding of reality. All I can say is... the people who have come out at night-time to the forest with us have, for the most, seen weird things and when I say weird things, I mean like orbs of lights shooting around the trees. So, Ricky and I are not alone in stating this... but what happened on Thursday evening (22nd. Feb) changes things for us. In essence a lot of this has already happened with my visits to Gordon's place and when he visited site 6 – but this was Ricky's and my own full encounters with the clan at site 6. Here is our report.

The night started off with Ricky and me setting up at our normal spot. There was a half moon in the sky with a crystal-clear sky with amazing clarity to the night sky – it was also very still and very cold. You could see very well through the forest darkness – with lots of shadows, but enough light to move around with – as long as you were careful where you were treading. Ricky decided to go for a wonder around site 6 by himself whilst I set up by playing various songs on Youtube for the clan. At one point I got touched on my shoulder, so I knew that my contact Zach was near. I started to mind speak with him and the answers I got was that the clan was near by and he himself was standing right behind me. I could just make out his shimmering image in places – but it is very hard. We didn't see any orbs – or at least none that we could both confirm as that. The reason is the moonlight made it difficult. In complete darkness you can easily see the orbs, but in this type of light it was tricky. I thought I had seen some, but I could not be sure. At one point Ricky crept up to me without my knowledge as I had no night sight as I was using the phone to pick out songs... he then snapped a big branch right next to me which made my heart miss a beat.... but if that had happened previously then I think I would have screamed... but these days, we are used to a lot of weird noises in the dark, so it was just a brief moment before I worked out it was just Ricky. So, joking aside, Ricky went off again to check the area out and all of a sudden I was greeted with Paul, come here quickly!! So, without torches, I made my way to see what Ricky was on about. And sure enough in this moonlit open clearing was a wave of shimmering shapes. A long line of them – all directly in front of us. Ricky could see them clear as day, whereas, I could only see them in parts – but they were there. It was the whole clan that had come along to greet us! Ricky and me were just overwhelmed by it... there – right in front of us – just feet in front was 7 to 10 of them making up a line of shimmering. There was one at least 9 foot tall (that was Zach). I started to mind speak to him and was told that the whole clan was here with us – and here was the crazy thing that I was told that 'they are with us and within us'. To us, that didn't make any sense... until later. Around this point, we both (separately) saw small shadows running around us at times – these were the juveniles and also Ricky got touched. Anyway, following this... Ricky walked around the group to establish if his eyes were not playing tricks and it was mist or something.... so following that Ricky heard a weird sound (which I missed) and then there were 2 amazing, loud clear whistles close by (within 100 feet) – these were no bird sounds. They sounded human but also mechanical in some way due to the pitch. It was amazing... and just at that point the shimmering disappeared. We waited a few minutes and then they were back again.... the connection between the whistles and the disappearance was interesting... but they did come back, which we acknowledged. We said our thanks and that we will be back again real soon. We then went off on a walk around the wood without the torches to see if we could find anything... which we did... but we ended up getting lost. That didn't last long and we got our bearings again and then went back to our spot, packed up and set off for home. It was only on the way home when we started talking about the night that it dawned on us what Zach had told me – 'we are with you and within you'. The car temperature gauge was registering minus 3. When I was in the forest I didn't have a hat on.... and yet, I was not cold. I didn't wear any gloves either and we both mentioned how warm we were feeling – the only exception being the tip of our noses – everywhere else was as warm as toast. It never occurred to us why this was. If that was any of our other trips there then guaranteed we would have been frozen solid – but that wasn't the case here and it was -3!!! That's when the penny dropped and we recalled Zach telling us that they were also within us! We cannot in any shape or form say that was the case- but why did Zach tell me that??? For me and Ricky, this was clear confirmation as to why we didn't feel at all cold that night. Too incredible for most people? Yes, it would be.... but for me and Ricky, we were there... we witnessed these events, we heard those whistles, we were touched by someone, I got that mind speak and the conclusions we had reached was that yes, indeed, they must have somehow controlled our body temperature that night so we didn't feel cold! That is in best the summary from the night. For me and Ricky this was the best night there ever. The clan showed themselves to us and we saw them... there right in front of us. Night time visits at home at 2.22 am ended the day I met Zach with Gordon – I now have regular mind speak with him. We have opened Pandora's Box to another world – and me and Ricky will be back again for more very soon. Ricky – care to add anything else in relation to that?'

15th. March, 2018

Thursday night, the 15th March, 2018. I was involved in an event that I am certain will cause a lot of back lash upon myself. I do not fear this, as I know in my heart and mind that what happened did actually take place. I witnessed and felt things that has been a first on my journey. When it comes to mind speak - all I can say is that I do get messages. How do I prove the things that I hear? Well, like others, I cannot do this easily. The way in which I get confirmation is by Zameath - Gordon's forest friend. Gordon cannot mind-speak with them yet, but he has a way of confirming yes and no to questions. If the answer is yes, then

Zameath reveals himself more. This has worked for weeks and weeks now and the answers I have received have been confirmed by both Mathew Johnson, Zorth and other clan members and the Exodus team. On Monday a report will be issued by Gordon about our meeting with 585 members of the Xanue. This comprised of our own clan members, the council of 12 including Zorth. And all the clan leaders from the clans of the UK. Ricky knows how to see the shimmering light that they give off, alas for me, I only see this occasionally or when they go a smoky grey colour. Gordon can see them very well, having been used to seeing Zameath every day. So, the importance of our meeting was to arrange a meeting with 'Patty' which Gordon arranged through Zameath and was confirmed by my contact Zach. My part came in in asking all the questions with Zach over a period of a couple of weeks and on the night. I was able to confirm, and I had to ask repeatedly for this, what Patty's name was..., and it was confirmed many, many, many times as Enrith. She is 147 earth years old too. She said that she would only partially uncloak on the night... which she did, going a smoky grey colour. Zach, and others and Enrith herself, confirmed to me it was her. It was a special night and I won't spoil the report that is coming out on Monday.... but all I can say is that it was a truly special occasion and I cannot back any of this other than mine and Gordons personal accounts. We did record with a Dictaphone the whole night, but not seen how good the recording is yet.... just for the record. I know that I will be victimised here, but I am more than ready to face it - because I know what I have been dealing with for so long now and this was the culmination of this. I was told that this was just stage one and that many more collective meetings will be taking place with us now. The next event will be at our own location.... and Ricky - you will be there for that one too.

24th. May, 2018

'This is a good indication of my forest friends trying to gain my attention. It's not been since November 13th, 2017 that they have visited me at night time and switched the TV on at 2.22 am. But that is exactly what happened last night. Several things are afoot that I am not telling right now that will literally blow people's minds.... but for now I have to keep tight lipped on it. I'll explain more in the future. Anyway, the forest people are making it known that things are getting a lot more interesting now. We have crossed some boundaries and there is no looking back. Right now, my path is quite a lonely one.... as it does seem impossible for people to believe the things I am saying... but they will have no choice eventually to take notice. It is reassuring to know that more and more fellow BF researchers are coming around to the woo side of things related to our forest friends.

Just bear in mind the name Enrith for the future!'

5th. June, 2018

Last night marked a huge milestone in my trips to the forest to engage in communications with the clan. I received mind-speak recently that they wanted me to go to the forest by myself last night, as I had been thinking about doing this for a while now.... so, having been told it was important for me to attend last night I said I would. Little did I know what challenge they had set for me. When I left, I noted that my lights were not working - so I had to drive home to change one of the headlight bulbs. After around 20 minutes I was on my way again... then to my horror I got to the junction with the main road to find it was closed - so I had to take a detour through the town centre to get to where I needed to go... I was starting to think that there were forces working against me this evening.... then when I got half way there... it started to rain! Even though it looked the perfect evening for the forest this rain came from no where...... I feel that I was either meant to turn home or I was being tested.... so I carried on regardless..... and sure enough - at the point I said... I'm still going regardless.... the rain stopped. It literally stopped at that point.... yet another coincidence? I just don't know anymore. Upon arriving at the forest, it was already practically dark - so, I took one deep breath and ventured off into the darkness alone - the place we go to is right in the middle of thick woodland up a long track. I set up for the night and said my thanks. What happened? I wasn't sure what to expect - but I felt quite relaxed. The one thing that I noticed far more than when I was with Ricky - was that being alone in pitch darkness - your hearing is so much more acute to forest sounds. They were there around me.... lots of footsteps and whispering in the background.... I saw 6 flashing orbs 40 feet in front of me...I was touched a couple of times and received plenty of mind speak. I was told that yes, it was a test for me - for me to be alone there in the forest despite other factors. I was there - alone with them right next to me - and I did not feel afraid. I played some music for them and then.... my mind-speak told me that I had been 'chosen'. Chosen??? For what I asked... and was told that will become apparent to me in the future and not to think about it right now. I know this is incredible - unbelievable - but this is the truth. I don't know what is being set up for me for the future.... or even if I am capable of doing what they are expecting.... but inside of me I felt honoured. I was told the whole clan and others were around me and they were all paying their respect to me.... How and why? Needless to say, my journey has been an incredible one to date from very humble beginnings....6 years ago. I know I cannot expect people to believe me..... but this is really happening to me! I have no idea how deep this rabbit hole is going..... but I am in total awe about it.'

21st. June, 2018

'It's been a while since this has happened.... but last night being the Summer Solstice.... I thought it was more than just a pure coincidence.... but last night I was visited again at..... you guessed it - at 2.22 am early this morning. The message I got was they understood my point of view on a question I had for them.... and they wanted me to come to the forest again... alone. So... that is the plan for this weekend! This is totally incredible... I feel my next visit will be a major turning point for me...... and hopefully I will have my new glasses, so I should be able to see them more easily from now on.. I'm really looking forward to my next visit.... and to think when Ricky and myself first started to visit the forest.... we dare not open the windows in the car when we were driving past the location - how things have changed!'

24th. June, 2018

'It was a first last night...... came downstairs to return some bits to the kitchen at around 9.30 pm (after the Germany game - which I was watching upstairs in our bedroom).... and then again I went downstairs to make a cup of tea at around 11 pm. To my shock... the TV was on by itself downstairs!!!!!!!! And it was the highlights of the Germany game! This was a first! No one had been downstairs between 9.30 and 11 pm.

No one could have switched the TV on downstairs... I asked if it was them and they said 'yes'. They wanted to let me know that a few of them were in the house.... and that made me smile. To anyone out there... they would certainly think I am a nut job.... but for me, I don't care... because I know the truth..... they are now my friends.'

3rd. July, 2018

'In the forest tonight to cheer on England.... and what a nail biting but perfect night! And now with my new glasses I was able to clearly see three nine to ten foot tall shimmering images standing right behind us watching the football game with us! Just a couple of feet standing right there alongside us. Now that is a first! We asked for their help when Columbia came back... and all I was told was 'to enjoy the game!!!!' How could I? It was going to end in a penalty shoot out.... but they already knew.'

3rd. July, 2018 – in the forest to see England play in the knock out stages. Talk about stress!

31st. August, 2018

'This time it was a visit to Gordon's site and we did a sleep over. It was a warm dry evening when we set up for the night and could see several shimmering figures surrounding us. Enrith was also invited and mind-speak confirmed that she was there. We settled down for the night this is when the most amazing thing happened. I was a bit uncomfortable as I had a trapped nerve in my shoulder, but I managed to fall asleep ok on the camp bed. Normally I don't have many dreams and certainly not real vivid dreams, but tonight was an exception. Half way through the night I had a surreal dream of being picked out of my camp bed and held the hand of one of the forest beings – Zach. I was looking down on myself as an 8 foot tall white lighted and hairy Zach pulled my soul from my body and walked me across to an open area and then spun me around, just like a cartoon character in a whirl wind, and then holding my hand we shot up into the sky at a tremendous speed and I could see the whole landscape below me…. I was then taken on a high speed tour of the world, but this was short lived as I awoke and could clearly remember what had just happened. I remember waking up feeling very relaxed and in awe. Initially I put this down to a wild dream, but it was so vivid, like no other dreams I have ever had. Anyway, I fell back to sleep again only for the exact same thing to happen to me again… pretty much exact. Zach spoke to me and confirmed that he wanted to show me things. When I awoke in the morning I was on cloud 9. It was the most amazing experience I have ever had with the forest people. I thanked Zach and asked if he would do this again for me sometime. The other thing is that my trapped nerve wasn't a problem again, so they must have treated me that night too.'

So what did I think had occurred that night? My best explanation was that this was some type of outer body experience and they had showed me just how they can travel the world at ease and with speed. This was incredible.

Chapter Nine

'Enrith'

Artistic reproduction of Enrith/Patty by Arfon Jones.

When I was developing abilities to mind speak with the forest folk, I had no idea where this would lead me. The first time it happened, that I cannot dispute, was when I was en-route to the forest and I was talking inside of my head trying to make contact. My question was, as you will appreciate, a profound question – 'What is the meaning of life?' Instantly, and I mean, instantly, I got a response in my head which I had no idea where it came from – it just wasn't from me. The answer was 'What is life?'. I was in shock. Where did that voice/thought come from. It wasn't me. I know it wasn't me… because it was a profound question in response, turning the question around. The rest of the day I was pondering upon this…. My question was asking for the meaning of life, but the response was… what is life? Profound question responded with another profound answer/question. As a result, it made me understand that to answer that initial question you need to define what life actually is. By our own current way of thinking, that meant some type of carbon based life form. And that is exactly what they were getting at. In order to answer the question, you need to understand the principles of what life is. Given how deep this first mind-speak was… where will this go?

Well, going on from there, I would ask thousands of questions and get just a handful of answers to them. It was basically on a need-to-know basis. I would always communicate with them, but getting answers to certain questions was impossible. Sometimes the questions would be answered with further questions or riddles, or something very different to what I had asked. Anyway, it was a year later that, given that I know that life can exist in very different forms, I asked again the question what is the meaning of life? This time the answer was just as profound and yet so deep that it makes you question yourself, as you are the only person to be able to answer it. They said to me: 'Does life have meaning for you?'. Maybe that's not the answer I was looking for, yet it was an answer and yet it was a question too. My mind was blown away by this and I thanked them for doing so. I think there is a lesson there for everyone, we are all searching for answers, yet the best person to answer them for us, is ourselves.

So, mind speak can be very challenging and sometimes very rewarding. It happens on their terms, and I respect that. If they were to give me answers to everything, then I would be asking for next week's lottery numbers. I understand it doesn't work like that. It's a need-to-know basis only. So, when Gordon asked me to find out the real name of Patty was (the famous Bigfoot from the 1967 Bluff Creek, California encounter), then I was all for it. I had found out the names of some of the clan members near me and also at Gordon's location… but this was on a different level now. On the 25th. of February, 2018 I was given the name 'Enrith' by my contact Zach

(Zachariah). I repeatedly asked for confirmation and each time I got the same response, yes, her name is Enrith. I was also told her age, in Earth years this worked out to be 147. I reported this back to Gordon and I had a quiet celebration with this knowledge.

It was a month later that I understood the significance of this. I was becoming aware of the fact that most of the names of the forest folk that I was making contact with had Hebrew names. Was this pure coincidence or was there any meaning behind this? Two of my own contacts were Zach and Sol (Solomon). I had looked up the meanings behind these two names to find their Hebrew origins and meanings... yet I hadn't done it for Enrith, as yet. So, I did. Enrith/Erith means 'Flower'. I double checked if it was Enrith or Erith and was told yes, it is the same, it's the modern English translation of the Hebrew name, but the meaning is the same. To now know that Enrith literally translates as 'Flower' blew me away... how wonderful is that? And it got me thinking.... When you visit someone.... what do you normally take as a gesture of goodwill? I have asked many people this question and I always got the same answer – Flowers! I think we can now reflect upon that 1967 encounter and see that the forest people allowed this to happen and the way in which they did this was to allow Enrith/Flower to be presented to us! Just how remarkable is that??

This is a sculpture that was done by the author back in 2015 and his understanding as to what Patty looked like – now known as Enrith.

Chapter Ten

Yet Another Year - 2019

There was a big pause in my reporting during this time. Basically, Ricky had other things going on and I had gone down a more personal route and going out on night-time visits with Gordon. We would do his site one time and then he would come over to site 6.

5th. January, 2019

'Back to the forest again today... and look who has been busy just 50 feet away from where we do our night sits....'

The author next to a large 'X' structure just 30 or so feet away by the night-time location. This was their sign. Both branches stuck in the ground making the 'X' structure.

16th. May, 2019

Site Report 15th. Of May, 2019

It had been a while since we had been to the forest, but tonight we organised a meet up. We agreed on a time and I was waiting for the message to collect Ricky.... but no message. OK, so I went to say goodbye to the wife, and she told me the internet wasn't working, which was correct. I then realised I had been waiting for Ricky's message. I turned on my mobile data to see he had sent me a message 5 minutes earlier that he was ready to be picked up at our normal spot. I said I was on my way... and I left. I got to the spot 5 minutes later only to find he wasn't there... I checked my phone and he said he was already walking to our house.... so... I set off to see him there. I never saw him going to collect him nor on the way back home. I got back home only to see that he had gone to the house, seen the car wasn't there and then started walking back to our pick-up spot again. This was crazy... so I called him, and we agreed to meet half-way between the two locations. This to me was a weird situation... call it a coincidence... but I certainly feel our forest friends had something to do with this.... anyway, we were now on our way. We reached the forest at about 8 pm (still very light and sunny) and parked where we normally say hello to our forest friends. They usually wait next to a tree - one or two of them, but tonight we could see around 20 of them - a crowd of them – all standing at different distances from us – from 8 feet away to 30 feet away. It was an incredible sight – all of them, as usual, in cloaked form. For me, it is normally difficult to see them as it is hard to distinguish them from the background – the best way to describe it is like a sheet of glass placed in front of you but with a very high vibrational heat haze or shimmer. In broad daylight it was so easy for me to see them today! And that was the start of a great night. We did our greetings and then went on to look around site 4 (the place where they place hundreds of ground sticks and stick structures). We asked them to come along with us as we did our tour and many of them certainly did so. I did pictures and videos of the things that we found and whilst I was videoing, we noticed one standing just to our right about 25 feet away. I talked out aloud to them and said 'I am videoing, I am going to start videoing next to you, please understand I mean no harm and you can leave if you like' and I gently turned the camera towards the one on our right hand side...... and shocked, totally shocked – the cloaked Xanue didn't move, didn't change in any form at all. Ricky and myself just stood there in awe at what was happening.... this was crazy.... and believe me, we thanked them with all our hearts. We pointed out where the Xanue was standing and how we could see it and how to describe how we could see him/her. There – in broad daylight standing 25 feet away from us. We celebrated at this milestone and achievement but acknowledged that knowing how difficult it is to see them with our own eyesight that they must already know that this would not show up on camera. I have not had chance to check... but be assured that will be my task after writing this report. Then again, we moved on and filmed different things and again I filmed another one.... this time I asked to come closer and I managed to get within 10 feet (still filming) before the shimmer disappeared. With mind-speak I said thank you and the answer I got in return was it was OK for me to film them... they did not mind! What an amazing evening – tonight was the best ever at site 4! and it wasn't even dark yet!

We then moved back to the car and got our night sit gear – just chairs and torches! No electronics other than our phones. We got to our spot at dusk knowing we would have a lot of moonlight for the night sit. We sat down and settled for the night – just talking for an hour as we let the darkness of the forest envelop us. It never really got completely dark as the moon was up and was three quarters illuminated. It was tranquil, the air was fresh and clean with a hint of pine and we were talking about why the air was so clear – maybe the trees were giving off more oxygen at this point? These are the sorts of conversations we have. We walked around to find a clearing where the moonlight flooded in..... we tried to take some nice atmospheric pictures of the trees in the moonlight. During this time we took loads and loads of pictures..... and yes, we noted there were moths flying around... but 2 pictures stood out as totally different to all of the others... these are included. Light anomalies are very common here... we see orbs all the time... but when the moon is out, we never see orbs as the moonlight drowns them out. If it is pitch black – we see them all the time when we turn up. So these pictures confirm something we have never achieved here before...... we then moved back to our chairs and settled down for a short while. After a few minutes we agreed that we would play some music to help raise the vibrational levels around us. We played a few songs and then talked out to our forest friends. We gazed around and could not see anything. Mind speak was quiet too, which was strange. Then Ricky saw a bright flash of light in the direction of a tree to the left of us, around 100 feet away from us. I was looking in a different direction at the time so I didn't see it. We stood up and was talking then all of a sudden Ricky shrieked out having a near panic attack.... he wanted the torch... saying over there.... over there..... I asked him to point out what he was looking at.... and he was holding on to me hiding behind my back for protection... he said it's coming this way!!!! (note: Ricky said it wasn't walking but gliding towards us). Ricky was so scared so I looked in the direction he was pointing at and there.... standing next to the tree – 100 feet away from us was a seven foot tall Xanue – in full physical form - partially silhouetted against the moonlit forest background. It was there standing there looking directly at us – and we were staring back at him/her. Ricky wanted to use the torch, but I said NO! So I shouted out hello! And tried to make conversations.... Ricky then dared me to walk up to him/her.... and rather than running away... I started to walk towards the figure (with no torches or electrical equipment). We got half way there and it disappeared. We carried on towards the spot and no signs – we could not explain this – it may have just disappeared into the tree!

We returned to our chairs and celebrated this monumental moment.... there in full physical form a Xanue was standing there with us both staring back looking at each other for a few minutes! Ricky asked if that was my guide and I asked with my mind-speak. The name I got was a different one to all the ones I had known of to date and I asked again and again. I was told the name was Morah. I then proceeded to check if that name was a Hebrew name (which is quite normal) and that is when the penny dropped! That name directly translated as 'the teacher'! Both Ricky and me were blown away! I believe Morah will be featuring in my life more from now on... but to what end? I am feeling amazed and humbled about what happened this night. After 7 years of returning to this location.... it was tonight where we came face to face, yes, at a distance, with a Xanue in full physical form. Our night was complete, and we look forward to the next milestone in our journey.'

15th. June, 2019

'Just for the record - Thursday/Friday am. Around 3 am (14th June). I woke up and saw an orb moving above our bed.... similar to how we see them in the forest. Just a short streak lasting a couple of seconds but clear enough for me to see it and recall it exactly in the morning. This is the first time I have clearly seen one in the house.... so that was something, but then the following day.... and Ricky will relate to this.... I have been hit yet again by the dizzy sensation that we get hit with after contact (sometimes). I still have it now and luckily it is not half as bad as it has been like in the past.... but this is them 100%. This is like the 5th time this has happened to me.... usually both Ricky and myself get this treatment at the same time. What does it mean? I have been told it is them reading my mind and also changing my vibrational level. They won't tell me why.... but I am certain this is for my benefit. I just wish they would hit me with the 'universal love' which they have done to Ricky and me a few times now. This sensation is like being on drugs - but your not, you are 100% in control - but life just feels perfect. I cannot explain it, it is just too crazy for words. But Ricky will agree with me that the dizziness is not a pleasant feeling.... but it does have some purpose for them. This might all relate to Morah entering my life I feel. Too crazy to believe.... yes.... you just need to experience it to believe it.'

3rd. August, 2019

'Lots of videos to add.... catching up on keeping a record of our activity. In this video you will see a 'clear' image of a Sasquatch standing directly in front of us... but when I say clear... I mean clear! But that is the stage we are at. More insightful videos to be added.... but this won't wash with people because they will not understand. The Sasquatch people are indeed inter dimensional travelling people. Take it or leave it... but that is the truth.'

22nd. October, 2019

'We see 100% of the visible electromagnetic spectrum. I suppose that's why they call it visible LOL. But we see less than 1% of the electromagnetic spectrum that light is a part of. The electromagnetic spectrum starts with radio and goes up in frequency through microwaves, then infrared, visible, ultraviolet, x-rays, ending with gamma rays. Cosmic rays are not electromagnetic, they are charged particles. The higher the frequency, the shorter the wavelength. All electromagnetic waves travel at the speed of light. (Copied from someone's post as I thought this would be of interest).

So.... bear this in mind when dealing with the Sasquatch forest people. Energies are the key....

That's it for 2019 – we were now about to enter Covid territory, so chances of getting out to the forest were very limited.

Chapter Eleven

Metaphysical

The word Meta comes from the Greek prefix of 'beyond' or 'after' and Physical basically being something 'tangible' (perceptible by touch) or 'concrete'. In the context that I am using this, is that what we are dealing with is something that is beyond the known tangible physical world that we know and accept or understand.

2020 itself was a very slow year for research. With the lock downs it was generally just me coming out to the forest to check upon any new activity and then recording this on video. But here are a few selected entries that I uploaded onto Project Zulu – the X Files.

23rd. May, 2020

'Second ever time I have had a stick thrown at my feet! Hand on heart, at that moment (and I would swear on my own life) - in my mind I was saying - you have left me no new features etc. I said it a few times over and then this happens! Yes, it is windy and could be just another coincidence - but it wasn't, and I was told so. It made me jump I can assure you, and for the record, despite the wind, I didn't see any other branches drop - a few leaves, but no branches - anywhere!'

6th. June, 2020

'Lovely walk around the forest this morning.... now I think I know why I was steered into going along this route - really beautiful and relaxing today.... I saw two of them who were following me. But the most amazing thing was when I got to our night time spot.... they had pushed in the ground a huge fresh branch - right in front of where we sit! Incredible. Obviously a sign - but they never tell me what it's all about. Pictures included.'

A huge branch/ground stick stuck in the ground right next to a tree at our night sit location.

Sometimes going Bigfoot researching is a real pleasure. This day was one of them.

12th. June, 2020

'Site report from the 9th of June to site 6 - this is Gordon's Report.

Met up with Paul last night and had an awesome meeting with the 4 clan leaders and their clans who we deal with. In addition, Enrith came over from Bluff Creek with 5 of her clan - 3 of whom had not been to see us before so gave a special welcome to them. In total there were over 50 Xanue present. As we sat and settled in the dusk, we could see the shimmering images grouping around us with Zac as host, standing 10 feet in front us - his 9 foot tall shimmering image easy to make out. Ison and Zameath as usual stood behind us. I played mostly Country & Western music for Enrith which she appreciated. We also talked about some of the Xanue aspects of life and as usual, some of the answers to our questions were cryptic! We lost track of time and it was suddenly midnight! The weather had been dry but it was then starting to feel cold. We said our thanks for another great get together and hope to have another later this year.

Additional Info.

Yes, we were surrounded by loads of shimmering figures and some were very close beside us. The reports from meeting up with Gordon over the last two years has been mostly recorded elsewhere. I have not been adding much here as my journey has been pretty much a personal one that has been far out there. I have tried to get the message across that we are dealing with intelligent inter-dimensional beings, and it has been a very slow process in gaining their trust - but I have no regrets about what I have done - for the most - people will not believe me, alas, I cannot do much about that. Mathew Johnson's new book will be out shortly and Gordon and myself have added testimonials to it. So watch this space!'

26th. July, 2020

'Just had this comment removed from a post off Bigfoot Believer's FB page - where they asked what you think BF was: "Inter dimensional shapeshifting beings of light. Those that know - know. Time for everyone else out there to have a bit more of an open mind about things." They only wanted people to say primitive ape! Oh dear - they still haven't made any of the connections yet and are still living in the dark ages....

As for Ricky and myself... we went to the forest on Friday night (24th July) and celebrated as this was the first time we had both been there together since the lock down. I had been with Gordon a few weeks ago.. but this was like getting back to normal. We saw a few cloaked figures whilst walking around and then surrounding us as it got dark. Both of us got touched during the evening... a bit of mind-speak confirming Ricky had a guardian (sounded like Mull for short or Mulion - a female - but I need to ask again about this for confirmation). Anyway... during the night the penny dropped on something very, very significant that happened during lock down that Ricky and myself have not spoken about. We intend to make a video regarding this as this is a story that certainly needs telling... but for now we are just going to list a record here as it seems recording times and dates is vital!'

This particular story dates back to Wednesday, 8th. April, 2020 – during the height of lockdown. It was crazy days back then. I want to reveal more, but I sense that this will have to wait for another day as it is too much for me, so others will discount it very quickly.

4th. August, 2020

Got my new research toy! I won't say any more here... but this will all be revealed in future chapters.

12th. August, 2020

'Nice walk around the forest today... not much has been happening here since I had last been..... but happy to see them 4 times following me on my walk! The last picture is a close-up picture of one of them - alas, no way of telling from a picture!' (see following page).

12th. August, 2020 – a 'crystal clear' picture of a Bigfoot
(standing on the left side of the tree here) – but in cloaked form!
No red circle around it, but what a clear picture!
As you can tell, it was for the author's eyes only.

THE BREAKTHROUGH

23rd. September, 2020

'23rd of September, 2020. This night marked a major step forward in our contact with the forest people. We deployed for the first time the gadget (my toy). The reaction to this was crystal clear - all three present witnessed orbs flying around the device. If we counted them, then we would have lost count, but we can clearly say way in excess of 50. Most were attracted to the electric field we had produced and stayed close to this - at about a foot away up to a metre - but we also saw some others around us in the tree canopy. I witnessed the brightest one I have ever seen here - at about 7 feet above Ricky's head and shooting off right above us into the trees. Some were short lived and not so bright, others were a few seconds long and bright. The success of tonight was incredible. Normally we would see occasional orbs, but these were scattered all over the place - this time we only needed to look at one single point and the forest people came to that point. Some all three of us saw, mostly Ricky and myself saw them at the same time and other times we each saw different ones at different times. It was truly amazing that we were able to have this level of visual interactions with them. Tonight history was made.... and going forward I am looking to increase the array of electrical fields to gain even better responses. Martin and Ricky... please add a comment to help confirm the above.'

20th. October, 2020: (from a different group – Xanue UK)

'Site Report from 20/10/2020

Gordon and myself ventured out to site 6 - getting to the site for around 7 pm (already dark with a crescent moon). The weather was very mild, still and dry. I noticed some very fresh activity with broken branches scattered around and a clear sign of a tree limb broken very noticeably where I would normally enter the forest from a path to get to our night time spot. Given the last visit, we were hoping for a similar memorable night surrounded by orbs. However, like we already know from repeated visits.... the interactions can change each time you come – it's never quite the same. That said, it is always pleasant to sit in the forest, in the darkness breathing in the pine filled oxygen rich air. There is something therapeutic about this. So, we sat down and had a chat for a while. Gordon could see a number of cloaked forest friends beside us.... and I sensed that there were many of them surrounding us, so I asked and got the response that there were over 40 with us. We spent some time catching up on things and we saw a few orbs – but nothing like last time. I saw only 5 orbs on the night – these were all very faint, but close by. We heard a few sounds, including possible footsteps.... but overall, a very different night to last time. We played a few songs and then packed up for the night and made our way home.'

23rd. March, 2021: (from a different group – Xanue UK)

'Site report from March 21st.

It was our first night time visit in 6 months thanks to the current lock down, but with the relaxing of the rules we were able to meet up and visit site 6. We, my friend and I, arrived just after it had turned dark, but there was a bright half moon which meant the woodland was fairly visible. It was mild, about 7 degrees and no wind. Recent forestry work had left a strong pine smell drifting through the air. We got to our spot and set up for the night. As it had been a while the two of us talked, given this was our first catch up in 6 months... we had a lot to talk about, and agreed on how much we had missed the peaceful atmosphere of the forest. We didn't see any of the forest people around us at the time... but we knew they were around us for sure. After a couple of hours I noticed a black figure run between two trees ahead of me about 150 feet away. It was about 6 foot tall and I discounted it being anything else despite how quick the sighting was. As the night progressed it started to get colder, so we got up and moved around a bit... it was at this point that I noticed two faint orbs pass by about 6 feet in front of me. I suddenly realised why we were not seeing many orbs during the evening as the brightness from the moon was drowning them all out. So lesson learnt... but given this was our first return to site 6 in 6 months, it was rewarded with another physical sighting. We have more plans for the coming months (if there is no third wave of this virus, that is). But it was nice to get back to the forest and spend time with our Xanue friends.'

16th. October, 2021: (from a different group – Xanue UK)

'Yesterday afternoon (bright sunny period) my wife came to me and said she was walking past the bathroom and from the corner of her eye she could see a tall grey shadowy figure standing there. She then walked back a couple of steps and it was gone, so she came and told me. We both went to inspect the bathroom (which at that time of day was the sunniest room in the house) and nothing was there. We did a recreation and I stood where she said she saw the figure and she said it was much bigger than me and around 7 feet tall. I asked (mind-speak) who it was, as I know they are always in the house, and was told it was Sol (Solomon). This was the first time that one of them showed themselves (in the house) as a shadowy figure, at least in daylight and to us.

However, our dog sees them (cloaked) as he always gazes at the corner of our room on my side of my bed... he sees them all the time and its fun watching him stare - as if at nothing. For me to see them in the house, it's tricky, and usually I just see them outside etc. Anyway, this episode, it shocked the wife but she accepts that we have them here (after all the other events we have been through here with them too). It doesn't worry her as she knows they mean no harm.... she doesn't share my interest in this but is happy to accept that they do exist, and we share our house with them. In many ways we feel very protected and very honoured. Can our Xanue journey get even more bizarre, well, yes.... it certainly does.'

25th. October, 2021

'Here is the picture from last night that has us thinking... is it or is it not? It was the first picture taken and it showed up here but nothing on the following pictures. It was right next to the device. It seems quite large yet diffuse but with a brighter starting point to the rest of the trail it made and looks circular like. When changing the contrast you can see a second one... again near to the device yet even fainter. I don't think this was a dust/pollen being blown by the wind as the directions of travel are going in different directions, plus I feel that the wind direction was going from west to east last night... again, different to the picture. And again zero on any other pictures. Why the first picture yet none on any others? I think we need to experiment taking pictures of insects at night and see how they look. Can we achieve the same sort of results.'

25th. October, 2021. Light anomalies surrounding the device (revealed later). What were they?

November 10th. 2021

'Today, 10th November, 2021 marks a date that the world has changed for me.... and I believe it will also end up changing the world for a lot of other people, especially in the BF world. I will be doing a video when I get the chance to confirm this... but this is amazing! Truly amazing! Sorry I cannot say anything more at this stage... I still need to join a few more dots in my research.... but I am right now blown away thinking about it all.... and I just needed to reference the date for my own records. Watch this space!'

22nd. November, 2021

'Just for the record. Last Monday I explained things to Ricky and his mind was blown away... the following day he found something that I had missed yet confirmed what I was saying but at the same time took away some of the news I was reporting, but not all of it. Yesterday I explained

A teaser.

things to Gordon and again his mind was blown away. They both understand why I cannot say anything at this moment in time.... but I am working on that. All I can say is that I have been looking at Sasquatch here in the UK.'

15th. December, 2021

'Short update: I have set up a meeting with a Film Producer/CEO of a reputable documentary film making company here in London. This won't be until the 6th (changed to the 11th) of January as he is away/too busy right now. I am certain this will end up blowing his mind too... but no idea if he will be able to run with it.... but feel this will be an opportunity he won't want to avoid. So, for now, I am keeping tight lipped on this as I want this to be released in the best way possible. I am so excited about this.... and maybe everyone in this group could be involved with this at a later stage???'

1st. January, 2022

Sad to see that the Forestry Commission has cleared a big area of the Goldilocks Zone. Very sad sight.

Part of the Goldilocks zone has been cleared by forestry work over the last few months. A very sad sight.

Due to seeing the above.... As we don't normally go here that often, as site 6 is on the other side and a lot of the tees that remained concealed what was going on, or being night time, we didn't venture there. So, during my mind-speak that night I apologised for the damage that had been caused and asked if that area was important to them. I got the response that the Goldilocks zone was called 'Samaria' – I remember writing this down following the mind-speak and it was only the following day that I looked up what Samaria meant in Hebrew and this made absolute sense. It translates as 'Watch Tower'. The Goldilocks zone was located at the highest point locally and had a huge clump of at least 50 year old pine trees within. I was sad that this important location had been partly destroyed. The forest people acknowledged my sadness and said that this happens all the time… and it will regrow again. My response was to buy a load of pine tree seedlings and go re-plant some of them myself. Given how small they were, I would rear them for a year until they are stronger and then plant them, yet 6 months later I was rewarded in seeing that the Forestry commission had already replanted the area and with much larger saplings. So, all is good, as long as they don't touch the other half of the Goldilocks zone.

Given the importance of Samaria to the forest people, I decided to also collect a lot of the left over off-cuts of wood that had been left by the forestry work. I hope that, in time, I will be able to make some useful memory mementos from these pieces in recognition of the area.

11th. January, 2022

'OK, job done… made my pitch and he summed up at the end of the call that he was shocked! He needed some time to think about this one and will be in touch again soon, possibly tomorrow. He had to go into another meeting after mine… so he works real late! Anyway, his only concern was that certain things I told him he needed a bit of extra confirmation for him.. which I explained would become available. He said it was a real great ending to his day but overall, he was in shock! Keeping everything crossed now…'

I never heard back from him, despite the reminders I had sent him. I was starting to get a glimpse of the reactions that I would get when dealing with the press and other relevant people. It was going to be total shock and closed doors from now on. And this is the exact reason why I am writing this book to get the message out there in the public domain. Part 2 of this book will explain everything to you. Unfortunately, I have had to rush writing this book and it has been put together within a 3 month period.

13th. January, 2022

'Report from 12th January, 2022. Trip to the forest, bottom end of site 4 (where the new x3 tree arches were). There was a 3/4 moon high in the sky, which was good for visibility of the forest, but made looking for orbs around the device being very difficult. I saw around 5 of them… very brief and dim. One of them both Ricky and I saw together... but Ricky was very lucky in witnessing an unusual and bright reddish orb. This was about 150 feet away and the size of a tennis ball... the picture below is Ricky's artistic representation. He also saw a bright flying orb in the tree canopy, but he wasn't sure if it could have been a shooting star... so this will need to be listed as a maybe. The most interesting thing that happened tonight was that, thanks to the moon, we could both see loads of shimmering figures surrounding us... just a few feet in front of us. They stayed for about 15 to 20 mins... and it wasn't any mist or the likes.. as it was a clear moonlit night and was -1 degrees and we didn't see any of these anywhere else. We talked about the meeting I had with the film producer yesterday and all of the crazy coincidences (?) associated with this... it was like this had been leading up to this years ago.... it was truly bizarre. Anyway, the cold got to our feet before anywhere else, so we called it a night and noted a few extra new tree/branch structures on our way back to the car. It was a good night and we wanted to show our appreciation for the direction things were going now.'

Bright red orb, the size of a tennis ball seen at site 4.

14th. March, 2022

'Quick video from the beginning of a trip we did to site 4 last night. It was a fairly flat night, but we had the opportunity to catch up upon our next steps regarding the big project, which has not been going the way I wanted.... but we have a plan B agreed (and also a plan C & D ready if required). We saw two groups of shimmering images at dusk near to where we park the car.... Ricky heard a big thud from a tree branch being thrown around a 100 feet away (which naturally I missed as I was talking), and we had electrical interference issues 10 mins. into the night sit... very unusual. I found out today that the device that stopped working last night was fine, but the 24v NEW adapter had fried! How that happened??? but this was at the same time as Ricky's phone being switched off. At the end of the night we went on a search around site 4 (as Ricky saw a dark figure (possibly a deer?) move ahead of us around 100 feet away... but there was no sign of it when we looked??? We then found a remarkable glyph.... but have no idea what it added up to... and this was just around 100 feet from our night-time sit (video to follow). And at the very end of the night... for the very, very, first time ever.... I tried doing a wood knock... so did Ricky.. but with no response. Of course. Overall, it was a very odd night, this may be due to us changing our normal night time sit location. Who knows?'

We never returned to this location again and always went back to our usual night time location in site 6. A bit more of a walk, but feel that if we keep following to our normal routine, then they will prefer this. Lesson learnt? Maybe.

10th. April, 2022

'Quick update from last night. It was a weird night! We arrived at around 7.30 pm and did a walk around site 4. We were greeted by strong shimmering figures at the usual spot. A video is included regarding the glyph/ground sticks combination we found in the dark last time and how bizarre it was. It had meaning in a language beyond our understanding. We highlighted the fact that there were more ground-sticks than trees at this part of site 4. We started to look around for shimmering figures and spotted one... so we took the unusual step of following in the direction they were.... would they lead us back to the car? Or take us somewhere different? We followed one of them about 4 times and took us to the boundary of site 4 that had a huge ditch and embankment. We found a broken off tree associated with a broad tree arch (pictures attached). No ground sticks here! The shimmering figures then disappeared.... but we found a nice place to camp over night in the future.

We then headed to our usual spot thinking we had upset them last time, given they fried my power controller to the device. We were laughing and joking heading up to our spot and watching some YouTube videos so we were totally unaware of what we walked past heading into the forest to our spot. We were greeted with two good sized ground-sticks right next to our night time spot. A good omen. We set up for the night and let the night time forest darkness envelop us. It was a very silent forest this night... very noticeable. No weird smells and overall, very still. We had a half moon above us which unfortunately drowned out any orb activity but gave us good sight around the forest. We caught up on things, played some music and had a few beers... a real good atmosphere. It was at the point that I was talking about the forest people having a guardian for each of us that something truly unique happened. Ricky saw a ghost like smoky mass (about 2 to 3 feet wide by a similar size in height) drift past our feet at about 7 feet in height.... right next to us! I never saw it as I was looking in a different direction... but it was described as something like breath when exhaled in the cold and just drifting by us both just a foot away! Naturally we checked that our own breath could not be mistaken and certainly wasn't as the air was too warm to achieve this. The excitement was soon followed by a huge fireball (meteorite) falling through the sky ahead of us... seen through the trees... but it was huge! The biggest meteorite I have seen in my life... the best way to describe it was like a flaming missile from a medieval catapult... not just a streak of light. It lasted a couple of seconds but burnt up in the atmosphere. Wow! I shortly afterwards saw an orb (which must have been fairly bright considering the moonlight) streak up a tree in the distance - about a couple of hundred feet away. So, a weird and very successful night... another unique closest approach we have had with one passing right next to us in visual form. Then... departing the forest we were totally, and I mean totally, blown away by the activity they had been up to right next to our entrance spot to the forest. Incredible and we have no explanations... but intend to come back in the daytime next to fully appreciate what they have done. Was it all natural? Natural and their involvement? Or just them? Well, based upon our initial assessment - it was them. They have done this before in the past at our entrance spot.... but this time they exceeded this a million percent! I'm still blown away thinking about it. Apologies if the video footage doesn't show up well... but you will get our impressions from our reactions. An amazing night!

One last thing... at the point we were checking the activity next to the entrance spot... we noticed a truly weird occurrence. The sky was crystal clear... not a cloud in the sky... yet from one horizon to the other there was a long thin streak of a cloud like structure. What was it! It didn't disperse or seem to move. Was it con trail at high altitude? If so, why didn't it shift? Well weird and we have no explanations for this. Pictures to follow!

Weird cloud formation on a clear night – going from one horizon to the next.

10th. April, 2022

I shared an article about energy, which I think is relevant.

'Too complex for me to understand… but the principle is there… energy can convert to matter and I would say, can reverse in the same way too. The forest people have mastered how to manipulate energy….. we are just catching up trying to understand.'

INVERSE.COM
Scientists managed to take pure energy and create matter — and new physics
When photons (particles of light; massless, pure energy) collide, they generate an electron a...

27th. May, 2022

'Report from last night - 26/27/May/2022. Basically, a dry night, with a little splattering of misty very slight drizzle - but essentially dry. No moon, but it didn't get dark till really late - after 10:30 pm. When we arrived, we scouted around the bottom of site 4 and saw that some kids had built a den - and they had used the sticks and branches of the ground sticks along the perimeter of site 4. That will be interesting to see what happens here... we did notice a few new ground sticks - but not on the direct border to site 4. Gem came along with us tonight as it's better weather and we were hopeful of a good evening ahead. Ricky and myself spotted the forest people waiting for us at the usual spot when we arrived and then we tested ourselves in site 4 to see if we could both spot one of them ahead of us (about a couple of hundred feet away), and sure enough - we both spotted the same thing - confirmation enough for us both. We then tried to ask Gem to see the same thing and, just like Magic Eye books, it takes a while for your eyes to adjust to seeing them... just like the green man - they are basically absorbed by the forest vegetation. As for the evening, we set up and found a new vantage spot to test next time we come - a deer stand. We checked it out and it gave us a commanding view of site 6. The reason we never used it before was that I was aware that a huge ground stick was placed next to our night-time spot pointing directly at it last year... so assumed that they were not happy with it being there. Slowly we let the forest darkness envelop us and waited to see orb activity - wanting Gem to see some etc. Well, just proves the point that we are dealing with an element that has a mind of their own... and the show never materialized. We did see some... but no where near the norm. Ricky saw a small red orb at the start.... and both Ricky and myself both witnessed about 5 at the same time... but they were so, so faint tonight. I must have witnessed a further 10 or so... but they were very, very faint. I did also see 2 sparks against the device. As for Gem... not being that interested, didn't see any.... however she has seen a really bright streaking orb at site 6 in the past. The funniest moment of the night was when we upset a Muntjac deer that had strayed our way... he was very vocal in displaying his displeasure... and this carried on for aprox. 20 mins... barking at us every 30 seconds or so.... that is... until Ricky shone a torch in it's direction.. then it would stop suddenly and all quiet... then start up again when the light was switched off... until... I made a thunderous clap with my hands and that was it... not to be heard again the whole night ! lol. We stayed until about 1:30 am.... in the hope that the obs will get more active... but not to be. Either way, it was a fantastic and fun night and to most people this would be an awesome night... alas for us, it was a bit disappointing, but still another lesson learnt, which maybe because there was three of us tonight and were too busy chatting rather than concentrating upon why we were there... who knows? It could be that we used the deer stand - which they objected to??'

30th. May, 2022

'Well, this was interesting... listening to Sasquatch Chronicles - Episode 845 'Corrections or Coincidences' it was a bit of an eye opener for me. For a very long time now, the reports from 1924 about Ape Canyon,

Mount St. Helens was used as the 'classic' story about a famous encounter with unhappy and aggressive Bigfoot. It was always told as an accepted confirmation about Bigfoot in the sense of flesh and blood animals. What I now understand is the whole paranormal side to the encounter and this was the first time I have ever heard anything about this! The story, if I wrote this down correctly, confirmed by Fred Beck, was that they had observed Orbs, had poltergeist activity and in the daytime they found single left footprints. Why wasn't this ever mentioned in all of those other reports about Ape Canyon encounter? Maybe because it doesn't quite fit the accepted narrative....

OK, trying to change that appears to be a very tricky thing to do too.... just like my 19th. Century hero found out, Gideon Mantell, but I can confirm that I have now put plan E in to action and will await the response. I can also confirm that I have invested in some more kit today that I really hope will have even more profound responses. I am really, really excited about this! Alas, running on a budget has not helped out here... but slowly we are making progress! Fingers crossed for this....'

4th. June, 2022

'Just a quick update from last night (3rd June). Weather overall good (some slight drizzle, but overall dry 95% of the evening. No moon, but didn't get dark until 11.00 ish).

Gordon and Gem came out to site 6 last night and a great start to the night before it got dark. Well, it was a brilliant opportunity for Gem as we were surrounded (within a few feet) of a whole host of shimmering figures - and Gem could see them this time! They were revealing themselves more to us than usual tonight. Every direction you looked we saw a line of them, including a very clear 9 footer next to the tree we sit next to. As it got darker, and we used the device we saw a few of them close to it. Again, the majority of the orbs were faint again... but again, Gem could see them too! We checked with each other the moment we saw some... be it a spark next to the device or a small flash. Gem has seen a really bright orb in the past, but this was tiny ones within a couple of feet of us! She was scared as well as impressed.... but don't think she will become a regular as the cold got to her as the night progressed. Gordon and myself saw two very bright orbs fly by just a short distance from Gordon... one flew in the shape of a semi circle arc. Very impressive. They both lasted about a second or two but very bright! Gem missed those ones... as falling asleep! Overall a very successful night and glad that Gem has now witnessed some of the things I have kept telling her about in the past.'

11th. July, 2022

Report from last night: 10th. July, 2022.

We set off early at around 5:45pm, reaching the forest just after 6pm. We were met by the usual crowd of around 3/4 shimmering images in the usual spot near where we park the car. We greeted them and then moved on into site 4 to take a look around together with a couple of cold beers given how hot it was still. Luckily, the moment you stepped into the forest glades it cooled the air temperature down by about 5 degrees. We commented upon how silent the forest was as we checked around for fresh evidence. We found new 10 foot tall (6 inch diameter) ground stick... which must have taken some effort to stick into the ground given how dry and hard the forest soil was. We then moved onto a big X feature. We were deciding if this was natural fall or made by our forest friends. We did a video on this... and due to the ground sticks being right next to it, we are inclined to think this was their work. Not the normal type of X, but from the Goldilocks zone and the bottom end of site 4 it looks very obvious. This X had appeared shortly after the forest work in the Goldilocks zone. It was here that the most unusual thing happened. A Red Admiral butterfly landed on my head... unusual.... but, when it flew off it kept coming back and landing on me. This carried on for 10 minutes... it was crazy, this has never happened to me in my life... and butterflies in the forest are rare to say the least. Ricky videoed 4 minutes of this. It only landed on me, and I can confirm I had no scent on. So, tonight was already showing some hope! It was back at the car I decided to use my new bit of kit - a full spectrum camera - and take a picture of the group of cloaked beings where we can see them waiting. I got the camera out and tried to take a picture... and at that point the batteries were drained! No way, these were new batteries... I could turn the camera on, but I couldn't take a picture. I tried changing the batteries and again, failed to operate the camera and at this point the group of them faded into the background. I gave up.

We then headed off to our usual spot at site 6 and set up for the night. All was good... so we thought! That would change.... as it started to approach dusk, we sat up in the deer stand so we had a commanding view over site 6. We did a 30 minute video looking back on the last 10 years that we have been coming out here and the progress we have made. Towards the end of the video Ricky saw a bright orb shoot across the tree canopy just ahead of us - 50 feet away. Another good omen we thought... we noticed that there was a three quarter bright moon coming up so we got down from the deer stand and got to the night time location. I set up everything and got it all switched on as normal. No problems. Half an hour into the night I noticed some weird things happening with one of the smaller devices...it was switching on and off. How? The feed is a constant power. Both Ricky and me saw a spark right next to it at one point... but orb activity was lacking overall. I had brought along a new bit of kit as well.... it follows the same principle of the devices and makes an electro magnetic pulse with rotating magnets. We heard some twig snaps at one point. We played some music and Ricky went off to investigate the twig snaps... and at the point I switched off the music, Ricky said a small twig was dropped on his shoe... but from where? Later on, I checked the devices and noted they were not working.... as it looks like one of the adapter units had fried (?). I swapped it over and sure enough, ten minutes later, it appeared fried too. This was working out to be an expensive night for me... so we said, back to basics and just sit and listen to the forest. I was already stating that I had a real tense head and the Ricky agreed that he was getting the same. At one point I heard whispering, Ricky had congested ears but at some points he could hear them too. We were now on edge like we had never been before.... not since we were first in the forest and plucking up our bravery.... we felt not welcome for the first time in 10 years! It was enough for us, so we packed up around 12:30. It was only when we got back to the car that behind us, from the direction we had just come... we heard and almighty crack of a solid thick branch being hit against a tree.... nothing like the tree knocks... this was a very, very powerful crack that we felt must have broken the branch being hit. We both understood what that meant... they were not happy with us..... so we apologised to them and made a quick exit. This is quite a turn around from all previous visits in the last 8 plus years. We understood that we had made mistakes and will not repeat them again. But that was quite an emotional overload last night!

The butterfly that kept landing on my head/shoulder directly next to the Huge X structure.

The following day:

> 'OK, well, luckily it appears that the full spectrum camera did mange to take a couple of pictures... so, did it work? This is, like all other pictures of Sasquatch... there is nothing conclusive... but I think the full spectrum camera has been able to capture a bit of contrast to where they are standing and the rest of the vegetation they were next to. They were not showing themselves very well at the time, so I think there might be some merit still in using the camera in future. They were standing next to the small tree.... I have, like all good bigfoot hunters produce when talking about pareidolia, have marked this with a red circle. But we can both see the shimmer when we are looking at them with our eyesight, but capturing this on film appears to be impossible... yet I feel this has caught something of a feature. Not great... but does coincide with the shape of what we would see etc.'

Full spectrum camera picture of the location where they normally stand greeting us on arrival.

The story behind this was that I had this camera especially for trying to capture the forest people, my new bit of kit to experiment with. It was working fine before we arrived, had new batteries and at the point I tried to take pictures of them, the camera kept failing. I kept trying…. And it would always switch itself off. So, I gave up… but luckily when I got home, I could see that I did manage to take a couple of pictures. Playing around with the contrast etc. I felt the black and white images showed off the distortion better, which you can just about see in the middle of the red circle here. Nothing to get excited about, but we feel that this was at least one way that we could capture at least something related to what we have been saying all along (that is, should they ever allow us to take the pictures next time?). Below is a full spectrum colour image.

Full spectrum colour image of the same location.

29th. August, 2022

'Report from Sunday, 28th August. We managed to get away early and got to the forest for about 6 pm. We were greeted in the usual spot by a small gathering of shimmering images, but not as strong as normal. We made our apologies for the last time we visited site 6 and then headed our way to the night-time sit. We sat down and had a chat over a drink etc. The mood was very different this time around. The weather was nice, dry and warm with little to no breeze. As we checked out the area, we found a load of wild raspberries and blackberries near to our night spot and tasty they were too. We got back to our chairs and surveyed the local area and sure enough we could both see that there were shimmering images next to the trees near us.... just a few feet away. They are never normally this close so early in the evening! Then we waited for it to get dark and after setting up the device Ricky was blown away... there was a smoky-distorted area right in front of his face... we checked if this was due to anything like the coffee steam... but this was nothing to do with this. It was also the same around me... but my eyesight was not as good as seeing this. Ricky had a tense head... not like a headache, but just pressure to his head and with a really happy feeling with it. After 15 mins the distortion disappeared. Then we started to notice very faint orbs. It was dark, but not dark enough to be pitch black. We checked if we were both experiencing the same things and sure enough we were both seeing the same faint orbs... one flew directly towards us... again, faint but we were both seeing them. Some were a quick sparkle and others were like mini shooting stars. Half way through the night I said I could see a visual distortion about 7 feet above the device, where the skyline was... Ricky then confirmed that he could see the same. So, how many orbs did we both see? We have no idea... but each of us saw in excess of 100 each. 20 to 25 we both confirmed that we saw the same thing... but the rest were seen by us individually so we could not say how many as so faint etc... but there were loads! We later saw the distortions again in the same place and we were hearing occasional twig snaps nearby too. Towards the end of the night we both witnessed a strange type of orb... it was wider... say an inch wide diffuse white light travelling from the upper left side of the device down towards the bottom left in a line. Very odd... we then played a couple of songs for them and then packed up for the night. No bad feelings tonight... far from it... all was happy. We got back to the car and no aggressive wood-knock this time! All is well again. One thing that we did notice on our way out was around the nest structure. This had been destroyed by forestry work over a year ago... and surprise, surprise... the only downed tree in the area was one pointing and touching the nest. So... just another coincidence? Anything you want to add Ricky?'

Given the feelings we had last time, we were happy that everything had returned to normal with them. The lesson learnt was no sitting in the deer stand (they object to these) and not to use the other device that I had brought with me last time. Another night with over 100 orbs seen!

22nd. October, 2022

'Report from tonight. October 22nd. 2022.

I arrived at site 4 at around 5 pm and was greeted by a line of shimmering images and then I had a look around the wider area by myself while it was still light. There appeared to be no fresh activity and the forest floor was covered in leaf litter. I then returned to the car to get the stuff to take on to site 6. Here I could see a very clear and noticeable shimmering image next to the tree I had set my chair up next to. Just a few feet away. I decided to create a much easier path into the site from the forest trail as Gordon was turning up a bit later. About half an hour later he arrived, and I met him walking up the forest trail. We set up two devices next to us and admired the host of shimmering images that were surrounding us. There were loads! It then started to get dark and sure enough the orb activity started. Gordon saw the first bright one shooting up the tree near him... then we both saw a lovely bright orb dancing around the device for a couple of seconds. Bright orbs were thin on the ground tonight, but as there was no moon, it was possible to see the faint ones. One that we both saw shot up a foot or so from the device next to me. We were then hit by some drizzle and later some heavier rain for about 20 mins. I wanted to see if the orbs still operated during rain, and they do! So, rain doesn't affect their electrical discharge. I also brought along a tuning fork set to 440 hz. There were various reasons for this and now I want to see if I get better response with the stronger device. Basically holding the fork next to the device I saw a spark right next to it... and this didn't come from the device!

After the rain stopped, I saw a real nice orb... and the best was to describe it was like a fairy flying around for a few seconds... yes, that is the best way to describe it - just like a fairy. It wasn't round... but looked like it had wings. Crazy!

Overall, it was another brilliant night. I saw at least 60 orbs on the night. Most faint. We then said our farewells.

When we finished and got back the cars, we were standing around and listening to an owl nearby, then I got a tap on my right shoulder from one of our forest folk! That was a real nice goodbye feeling!'

27th. February, 2023

'Report from last night: 26th. Feb, 2023.

I went to collect Ricky from his place, about a mile away, for 7.30 pm. On route the weirdest thing happened in that the roof of the car was being banged on whilst driving. It didn't happen just once, but twice (half a mile apart). I cannot explain it... but when I got to Ricky I checked the car over and nothing, explanation wise. At the end of the night Ricky did a re-creation of banging on the roof... from outside with the doors shut and the engine running. We did several recreations until I could say that it was just like that, and it was right at the front of the roof on the passenger side and like someone knocking... that is the best explanation I can give! Well weird...

We got to the forest and set up for the night. It had been a while, so we had a lot of catching up to do. It was very cold, just above freezing point, but the conditions, apart from the semi moon, were perfect. We stayed about 3 hours and towards the end we started to play some music and that's when I started to see the odd Orb around the device. The moon was preventing most from being seen well, but we saw a few every now and then... and after a while we both stood around the device... just 12 inches away from it... that close... and low and behold, we both started seeing orbs in between us and the device... so literally just 8 inches away!!! Incredible. We must have seen around 20 of them in close proximity to us...

We then said our goodbyes and made our way out of the forest and at this point we noted the most amazing ground stick ever! What a nice present they had left us. The ground stick looked amazing... very different to any we have ever seen in the forest... and it was pointing directly at the stump that we put the device on... at around 40 feet away from it. The stick itself was covered from top to bottom with bright silver lichen. It looked incredible and the pictures don't do it any justice. This was certainly a thank you message pointing at where we use the device. How amazing was that!'

This is the latest update to Project Zulu – X Files prior to organising the publication of this book. As you can tell... we are encountering even more unusual situations that we have never seen before.... And we just love it!

Magnificent ground stick covered in silver lichen left for us at the Night-time location, February 26th, 2023.

Something to consider:

String Theory proposes the idea of higher dimensions and parallel Universes. These parallel universes might be a fraction of a distance away from us, indeed, overlapping us, but we just don't sense their existence because they are vibrating out of phase with our own physical world.

This is the end of Part 1. The entries that I have shown here just gives you a brief snapshot of what we have been up to over the years. The hours and hours spent investigating and researching was paying off slowly, but surely. Most of this was a personal journey and carried on behind closed doors. People were very hostile towards what I was saying, yet deep down I knew that the evidence was taking me to a place nobody wanted to go to. All of this led to the remarkable discovery which will be revealed in Part 2.

Recent picture of part of the Goldilocks zone – December 2022.

PART 2

Merrie Olde England!

The Mid Winter Festivals of old....

Today, as a Christian nation, we celebrate Christmas (the Mass celebration of Christ) on December 25th and we recognise this as the date that Jesus was born in Bethlehem. We know this and accept this, it is ingrained into our traditions but where did this start and do we have any supporting evidence to confirm that this happened on December 25th? Well, I don't want to get into any religious debates about this, but as far as I am aware, there is no actual evidence supporting this.... even the year that this took place is heavily debated. So, can we trace some of the origins to where we got this date? Well, yes, I believe we can. This happened in the reign of the Roman Emperor Constantine, in the year 336.

Constantine I (The Great) Bronze Follis coin – circa: 307 to 337 AD – depicting Sol – the Pagan Sun God (prior to the change in the Roman religion).

Constantine converted the Roman Empire to Christianity – a major achievement back then and there must have been good reason for doing so. Prior to this, the Romans had various mid-winter celebrations – Saturnalia (The God of Saturn – and the god of agricultural bounty – celebrated from the 17th of December through to the 23rd of December (Julian calendar)) and more importantly – 'Sol Invictus' – birth of the Sun God! This was celebrated on, you probably guessed it, December 25th. These were all considered Pagan festivals, and these were to become outlawed, and the new Christian celebrations were then to be followed. Why was Sol Invictus so important as a celebration all those years ago? Indeed, why were mid-winter celebrations had around the pre-Christian world? All I can say is – look to Stonehenge. The pathway of the sun along its yearly celestial trail was of great importance – when to plant crops, when to harvest – it was a seasonal calendar. So, the Winter Solstice was good reason to celebrate as the sun had travelled to its most south easterly sunrise point before heading in the opposite direction. Solstice basically means the 'sun standing still' – which happens at both the summer and winter solstices – given that it appears that the sun stays in the same location for a few days before reversing its direction along the horizon at sunrise/sunset.

So, just like the early neolithic farmers on Salisbury Plain, watching the movements of the sun was vital and as a result, Stonehenge was a means to track the sun (and maybe the moon as well) across those Wiltshire skies all those thousands of years ago! Celebrations still continue at Stonehenge on the Solstices each year as testament to this – I too attended one of those occasions. For me, it was very special…. being able to witness the sun rising above the heel stone on the Summer Solstice – and to the builder's that built this incredible monument 5000 years ago, you can only imagine just how important the Solstices' were. Indeed, the mid-winter solstice would have been far more important to these early people, as it spells the turning point of the winter months. This is where we get the term 'Yuletide' – meaning the 'time of the solstice'.

So, the significance of the Roman feast of Saturnalia and Sol Invictus were directly associated with the movements of the sun and Sol Invictus was the day that the Sun God was 'reborn' – what Constantine did was to use Sol Invictus as the date of the birth of the son of God! This kept the pagan festival alive, but changed the emphasis of what the celebration

The sun rising above the heel stone at Stonehenge on the summer solstice, from when the author attended.

represented. To this date, December 25th has stuck, but it is interesting how religion and past beliefs have morphed into the festivities that we know today. The research I did into this has been a real education for me and I hope that you will be able to understand this and double check certain things for yourself. What I am confirming here is a pure summary of this, and some parts may, or may not, be subject to debate, so this is my own interpretations.

Now, we are starting to understand that our Christmas period has a direct link to previous Pagan festivities celebrating the change in direction the sun makes along the horizon. The daylight hours are going to get longer now! That would have been a good reason to party and that is exactly what they did – including pre-Constantine Roman times. Saturnalia had up to 12 days of festivities. It is interesting to note that there are 12 days between the winter solstice (December 21st) and the new year. And the word January comes from the Roman God Janus – which was the God of new beginnings or transitions. Interesting don't you think? Could this be where we got the first 12 days of Christmas from?

Was there a link between the birth of the 'Sun God' and/or the 'Son of God', I am sure there is, yet my knowledge regarding this time period is limited and I am sure that other academics and religious scholars would like to comment on this. For me, this is just setting the scene.

Introducing Robin:

In November 2021, I was watching a programme about the potential origins of Robin Hood, as I had feelings that there may have been some connections to my own research and during this programme a character called Robin Goodfellow was mentioned. So, this was the turning point for me to look for material related to this medieval character. It was on the 10th of November that the penny dropped for me upon receipt of a copy of a pamphlet written in 1628 - 'The Mad Pranks and Merry Jests of Robin Goodfellow'. This was a copy of a singular pamphlet that had survived the test of time that was written about Robin's folklore antics. These types of pamphlets were popular for their time, but being purely printed on sheets of paper, they rarely survived. We only know of this copy from 1628 and a later second edition from 1639, however you will come to appreciate that Shakespeare borrowed and adapted this folklore character to have a starring role in one of his plays back in 1595/6. As a result, we know that Robin was known about prior to 1595.

Now, is Robin Goodfellow a fictional character or is there some truth behind who he represents? To scholars, I am sure that this character is nothing more than a larger than life, fictional folklore individual from Medieval England. That said, we appreciate the value to some folklore stories - many are based upon stories and rumours of the local people. This is nicely confirmed by the stories surrounding mountain gorillas leading to their eventual discovery in 1902. Lessons learnt? Should we then dismiss the merit of stories from the past…. Or is there some real value there? Let's explore that.

First of all, I am sure that Robin Goodfellow would have been a household name back in medieval England and openly talked about in the taverns of the day. Although we don't know the original source of this character, we do have some pointers to who he represents. Shakespeare had portrayed him as a magical woodland sprite with abilities beyond people's understandings yet having good intentions. In the play he was called Puck, but also known as Robin Goodfellow. The name itself, Robin Goodfellow, is a bit of a giveaway - as he was believed to have been a good fellow! Now, Shakespeare's play is just that…. a drama, and a drama based upon a summer's dream, so there is no value to its basis…. Or is there?

Well, thanks to my past 10 years of research, the hidden meanings behind some of the content within Robin Goodfellow's 1628 pamphlet hit me like a sledgehammer! And as a result, this just snowballed into a much bigger idea, the likes of which would challenge our modern day accepted understandings, traditions and will indeed, rock the status quo. The truth I have uncovered had to be told!

Having read earlier chapters, you will understand the unique nature to the forest people that I have been interacting with. So, quoting extracts from the pamphlet you will most likely understand the significance:

> 'Some call him Robin Good-fellow,
> Hob Goblin, or mad Crisp,
> And some againe doe tearme him oft
> By the name of <u>Will the Wispe</u>;
> But call him by what name you list,
> I have studied on my pillow,
> I think the best name he deserves
> Is Robin the Good Fellow.'

And,

> Get you home, you merry lads :
> Tell your mammies and your dads,
> And all those that newes desire,
> How you saw a walking fire.
> Wenches that doe smile and lispe
> Ues to call me <u>Willy Wispe</u>........

You will see that I have highlighted the words 'Will the Wispe' as we are totally aware of the orb activity in the forest. This was a clear pointer to Robin being related to these forest beings - in deed, Shakespeare was also well aware of this too - by making Puck a forest sprite and also his attempting to do good deeds. Everything that followed on from this just confirmed things for me, and I will quote another part of the pamphlet that will stick out too:

> 'I have the ability to change thy shape,
> To horse, to hog, to dog, <u>to ape</u>,
> Transformed thus, by any meanes'

Again, I have highlighted 'ape' here. For Medieval England to quote 'ape' is quite exceptional. Why would this be? Would this be based upon pure fantasy, or would it relate to actual sightings? Well, it doesn't take an expert to put two and two together here, now does it?

We are now looking at Robin Goodfellow being associated with orbs and being a shape changer, morphing into ape like beings. Brilliant! In Merrie Olde England, they were trying to make sense of a known phonon omen, and the best way of doing this is making inflated stories about them, including Shakespeare! How incredible is this? Shakespeare was writing about Bigfoot all those years ago.... And nobody knew. Blown away yet?

Ok, so that is a bombshell in its own right.... But the significance of Robin Goodfellow does not end there. The next bombshell is coming up next....

The following quote from the pamphlet will be an absolute give away:

> 'My nightly businesse I have told,
> To play these trickles I use of old:
> When candles burne both blue and dim,
> Old folks will say, Here's fairy Grim,
> More trickles then these I use to doe:
> Hereat cry'd Robin, Ho, ho, hoh!

There is a slight change to the modern-day quote of 'ho, ho, ho', yet we know who is the only person around the world with that one liner.... And you've guessed it.... Father Christmas - Santa! Oh my word, dare I think this? This stamps upon the accepted ideas as to the origins of Santa - the 3rd century Saint from Turkey. So, what do we know about Robin Goodfellow being related to this larger-than-life character that visits us on Christmas Eve? Again, the pamphlet from 1628 helps confirm all the things that we know about Santa and yet there is zero evidence behind the accepted Saint Nicholas origins to Santa.

We all know that on Christmas Eve we leave out milk and cookies as a thank you for Santa visiting us, but where does that idea come from? Well, in medieval times people would leave out bread and milk for the fairies in order to do good deeds (gifting), to bring them luck or stop them from playing tricks on them. And here is a quote from the pamphlet:

> 'Because thou lay'st me himpen, hampen,
> I will neither bolt nor stampen:
> 'Tis not your garments new or old
> That Robin loves: I feele no cold.
> Had you left me milke or creame,
> You should have had a pleasing dreame:
> Because you left no drop or crum,
> Robin never more will come.
> So went he away laughing ho, ho, hoh!'

It has to be emphasised that upon departing, Robin would call out ho, ho, hoh! This is quoted many times in the pamphlet. As a result of not leaving out milk and bread, Robin would not return to this household. So, this is the basis behind our modern-day tradition of gifting on Christmas Eve. It's a 500 year old tradition!

Now, how did Robin get into the house? Well, the belief was that he came down the chimney as there was no other way that he could get into a locked home. We see references to this in early images of Robin carrying a chimney sweep and here is another quote from him:

> 'Black as I from head to foot,
> And all doth come by chimney soot'

Robin Goodfellow, as portrayed in the Mad Pranks and Merry Jests of Robin Goodfellow, 1639.

So, is this enough to point towards this Medieval English character being the true origins of Santa? Well, I believe so, but let's go further with this.

Robin Goodfellow as the Wildman.

The earliest references of Robin Goodfellow were that he was 'elf' like, he had a holly crown and ivy as a beard. In fact, he was the undeniable depiction of the Green Man, and we are all familiar with images of the green man with foliage sprouting from his face. These are all well represented in British Medieval Churches and Cathedrals. The Green Man is a man of the forest hidden by the forest - and we know how that works with the forest people and their incredible abilities.

The Green Man. A reproduction of a carving from the medieval choir stalls from Beverley Minster.

The same applies to early references to Santa being elf like and only in 1886 was he drawn as full sized and human like (in the manner that we recognise him today) by the artist Thomas Nash. Prior to this he was represented as part human/part elf, but gradually becoming more human like and you can appreciate how he looks today. Over time, as people's ideas change, so did our representation of Santa change. Indeed, as recently as 1931, Coca-Cola, as part of a promotional idea to get more people drinking coke in the winter (given it was considered more of a summer drink) got Santa wearing a bright red coat and advertising has enforced this ever since. Basically, he was rebranded…. Yet the original Father Christmas wore mainly green (the Green Man), although some later representations were shown in different colours prior to Coca Cola. You can see this in Charles Dickens Christmas Carol where the ghost of Christmas Present, in fact, an early representation of Father Christmas, is dressed in green, wearing a holly crown and holding a fiery torch. This was how he was represented in 1843.

Charles Dickens Christmas Carol – The Ghost of Christmas Present – 1843.

There is also a link here to some medieval coins that I came across representing the green man from Germany.

You can see that the Green Man is hairy, wearing a crown of vegetation and carrying a fire torch or 'wisp'. So similar in many ways to the way that the ghost of Christmas Present is represented. This coin dates to 1595, prior to the Mad Pranks of Robin Goodfellow. So, is this one of the earliest representations of who was to eventually become Father Christmas?

Postcard from 1910: Here you can see a green Santa sitting beside the fireplace. You will also see Christmastide within the wording. Just like Yuletide (Time of the solstice), Christmastide was the time of the mass celebration of Christ. We now just say Christmas time, but it didn't always use to be like that. Yes... things change over time!

Here we can also see a picture of Father Christmas from 1910, dressed in green. So, do we now have a link as to why Santa lives with elves? I think it is very clear about everything that we accepted related to Santa is confirmed by the references to Robin Goodfellow. Now you can imagine how this will cause a stir to our modern-day beliefs and traditions? We have accepted Saint Nicholas as the real Santa, but he is not. He has been made to replace previous held beliefs about the magical forest sprite - Robin Goodfellow. We can now appreciate why Santa is so magical, why he can fly around the mid-winter skies. We can also make reference as to why we put a fairy (not an angel) at the top of the Christmas Tree. Please also note the tradition of a Christmas tree you might think dates back to Victorian times and having been introduced by Prince Albert from a German tradition, but the concept goes back much further in history, including the UK too. It was customary to bring in ever green foliage (Holly, Ivy and fir branches etc.) into the house as a good omen for the coming new year or to ward off evil spirits. As a result, these plants were believed to have magical abilities… mainly because they remained green in the winter when all other plants died off. So, the reason behind the Christmas Tree dates back to our early mid-winter celebrations, it also represents the forest where Robin comes from, and we make our offerings of milk and cookies in front of that very tree. It is also worth noting that

Who is it associated with?	Green Man / Robin Goodfellow	Saint Nicholas
Giving gifts to people	✓	✓
Origins prior to Christianity	✓	X
The reason why we have a (Christmas) tree in the house at this time of year	✓	X
The reason why we have Fairies (not just angels) hung on the tree (including fairy lights)	✓	X
The reason how Santa can fly around the world magically	✓	X
The reason why Santa uses forest animals (deer) to pull him	✓	X
The reason why Santa comes down the chimney	✓	X
The reason why we put milk and cookies (bread, in historical context)	✓	X
The reason why Santa is 'magical'	✓	X
The reason why Santa lives with elves	✓	X
The name Santa has links to the person in historical reference	✓	X
The reason behind a naughty/nice list	✓	?
Origins of being dressed in green prior to Coca Cola	✓	X
Why Santa is always 'Merry' and says 'Merry Christmas!'	✓	X
The reason why Santa goes Ho Ho Ho!	✓	X

Quick overview of the accepted and the now believed origins of Santa

Robin describes himself in one word as 'Merry'. It all makes sense now. I have included a table that tries to link the things that we know about Santa and how it relates to both Robin Goodfellow and Saint Nicholas. You will see, there is no contest!

One last thing that we all want to know: Santa's naughty and nice list. Is this real or not real. Well, it has to be said that this has been altered over time in order to become more in keeping with only receiving gifts if you are deserving, yet the pamphlet also links in with this, as Robin Goodfellow was renowned for playing tricks on people who were bad and yet doing good deeds for those that were good to him. Here is a passage from the 1628 pamphlet:

> 'But to the good I doe no harme,
> But cover them, and keepe them warme:
> Sluts and Slovens I doe pinch,
> And make them in their beds to winch.
> This is my practice, and my trade;
> Many have I cleanely made.'

So, in summary, you can see that Father Christmas/Santa can easily be linked to Robin Goodfellow, yet everything related to Saint Nicholas can be discounted on the same basis. Just like the decision to create Christmas Day on the 25th of December to supersede the Mid Winter Solstice celebrations, a sense of religious propaganda was used to make reference to Saint Nicholas representing this pagan character from medieval England.

Let's now look at the name 'Santa'. This is where things get more complex and the cross over to his name becomes shrouded by the mists of time. What we do know is that the Church outlawed all Pagan celebrations in the middle ages, this became especially significant during the Puritan period following the English Civil War (1642 - 1651) where Cromwell, Lord Protector, banned all plays and any recognition of Catholicism, Pagan spirits, and possibly even banning Christmas celebrations (it is understood that the Puritans didn't celebrate Christmas as they didn't believe in Christ being born on December 25th). Britain itself became a very different place, more strict, less merry and it was only following Cromwell's death and the restoration of the monarchy with King Charles II that things began to change back again. Britain's celebrations continued with the birth of Christ together with the giving of gifts represented by the three wise men. However, other parts of Europe continued with their pagan beliefs with a 'Sinter Klass' travelling about giving gifts mid winter. In 1821 an illustrated anonymous poem called 'The Children's Friend' had represented Sinter Klass dressed in furs with reindeer pulling his sleigh. The term Father Christmas first appeared in 1616, when it was mentioned by the playwright Ben Jonson in his play: Christmas - His Masque. It was a performance for the Royal Court of King Charles I. It was a folklore play all surrounding Christmas, although his appearance was more in keeping with the Jacobean era. The name Santa also has a connotation, as the word Santa is also an anagram of the word Satan. Horrifying as this is, but we have to remember that when the Church outlawed the Pagan Green Man, they termed him as Satan. So, there is a link here, but is it a correct link? I don't know, but you will see that over time, elements of different cultures and also religion have shaped the development of Santa Claus. It is fair to say that the majority of this comes from Robin Goodfellow, otherwise we would not have the classic saying of Ho, Ho, Ho or be leaving out milk and cookies for him on Christmas Eve!

It is also interesting to note that the name Robin/Robert also had a nickname of 'Hob' in the Middle Ages, the modern term is now called 'Bob'. As you will also appreciate that Hob was a short name for Hobgoblin… which also referred to the Devil.

Now, where does this leave us? If we are to say that the true origins of the magical Father Christmas are Robin Goodfellow and Robin is based upon the Green Man of England, then, I'm going to make a bold statement that Santa is no myth, he is real, and his spirit is still alive to this very day! No more having to tell children when they get older that he is not real - because, in essence, he still is alive to this very day. What a fantastic reason to celebrate! The magic of Christmas, from when we were children, has come back to life! This has affected me greatly in a very positive way, and I hope this will have immense positive effects on everyone else. Naturally the way in which we perceive him has changed remarkably since Medieval England times, but from our own experiences we know that these forest folk are good people. They are intelligent people, and they are magical people!

For the record, I came to all these conclusions myself, but based upon subsequent research, I was able to add extra validity to these claims. I was not the first person to make the connection between the 'Wildman' and 'Santa'. A lady by the name of Phyllis Siefker wrote a book back in 1997 – Santa Claus, Last of the Wild Men. She spent 10 years researching the origins of Santa – from 50,000 years ago to the way we accept Santa today. Unfortunately, she has passed away now, as I am sure she would have loved to see how things have progressed. Her work was very detailed and there was a good chapter covering Merrie Olde England, but she missed some of the important points about Robin Goodfellow. Her conclusions also missed the root source as she understood that the wild men were based upon a bygone civilization, and she based her thoughts upon an isolated northern Japanese culture as an example – the Ainu people (considered stone aged culture). These people were eventually absorbed into a more mainstream Japanese civilization (yet, taking up to 1789 to achieve) and as a result, they then had to give up their more traditional lifestyles which was more linked to the natural world and the spirit world. So, Phyllis was correct in her beliefs in Santa being linked to the wild man, but falling short of who this actually related to. I am pleased that she had spent all those hours researching this as it has helped validate my own research and I would like to acknowledge her work. I would also encourage others to obtain a copy of this book to gain a more in depth look at the stories relating to the emergence of Santa. It would also be beneficial to get hold of a copy of 'The Mad Pranks and Merry Jests of Robin Goodfellow'. That way, you now have the historical background to the links.... later on, we will look at ways that you can add your own validation to the claim that – in essence – the spirit regarding the origin of Santa is still very much alive today.... together with all the perceived magic and association with fairies!

As for Father Christmas, Santa, going forward... I now feel that we should rebrand him once again and this would be an Emerald Green Santa, keeping to tradition and now knowing his true origins and that he still lives around us, and within us, to this very day!

Just another coincidence

Years before I made the above discovery, or even knew about Enrith, my mind was preoccupied with Patty. So, ironic that it is, I had made my own sculpture Xmas tree decoration featuring Patty. I was also fortunate enough to know somebody going to the USA for the Salt Fork Bigfoot conference which Bob Gimlin was also attending. He was able to give Bob one of these tree decorations and also get another one signed for me. Years later, I am blown away thinking about how much of a coincidence this would eventually become!

The Xmas tree decoration made for Bob Gimlin to sign.

Signed on the reverse by the man who witnessed Patty/Enrith back in October 1967.

The Green Man
I do believe in Fairies, I do, I do!

The history related to the Green Man is complex and covers many different countries and beliefs, so it is impossible for me to do this any justice in this book. What we do know is that some of the earliest origins come from Greek mythology and the God Pan (God of the Wild). Pan had the hindquarters, legs and horns of a goat, very much in keeping with the later descriptions of Satan, however this was not the manner in which he was worshiped. It is also worth noting that the author of Peter Pan, James Mathew Barrie, had based Peter on this mystical Greek God (written in 1904). So, with the connection of the fairies, Tinker bell, and another world – Neverland, made this fairy tale is a bit closer to the truth than we can dare to imagine. I wonder how James Barrie came up with this idea.

Greek Coin depicting the Greek God 'Pan' – circa. 310 – 303 BC.

We all know that the 'Green Man' was not restricted to Greek Mythology... the Romans had the God Faunus (God of the animals) – which basically mirrored Pan, they also had the Roman God Silvanus – God of the forest and the countryside. There is the Celtic God Cenunnos or Carnonos, who was another horned god, but with antlers and believed to be the God of animals. It is interesting to note that the relevance of the horns was related to the power of 'enlightenment' rather than anything being classed as evil. Each of these Gods alone could have a thousand words written about them, but I am just giving a very brief outline here as the focus is the Green Man of the United Kingdom.

It is well known in Britain that early Churches built from the Norman Times through to the Middle Ages would have sculptures of the Green Man. They could be hidden within the roofs and gables of the church, inside or outside or more prominently displayed. We see the same images portrayed with a face hidden behind dense foliage, with some leaves or branches sprouting from the mouth, nose or tear ducts of the character.

This fine green man is located at Sutton Benger, Wiltshire.
Age unknown, but is believed to have been re-touched following an earlier carving of the Green Man. There are other Green Man carvings around this church, which certainly appear to be much older. See later.

Why then would the Church include such symbols? Well, I think we can see that there has been an overlap with Christian and Pagan beliefs. They have morphed over time, but back in the Middle Ages, people were more connected to the natural world and certain pagan beliefs. I am guessing that in an attempt to include the Pagan Gods of nature, then the Green Man being represented in Churches would be a fitting way to combine the religions. And this is what we have also seen within the Christian celebrations regarding Yuletide – but it doesn't end there, more of that to follow later. The Green Man is clearly a case of symbolism – the old and the new. Acceptance of the old yet ushering in the new religion – it basically represented a bridge! It is also said that they were there to ward off bad spirits. The symbol of the Green Man was that powerful, that most early Churches had him represented.

So, what does the Green Man represent? Being green, even in the modern world, represents 'new' as well as just the colour. Think about it, when you are new to something you are considered 'green'. The Green Man celebrations was all about new growth... be it Father Christmas bringing in the Yuletide celebrations or in the Spring – when there is fresh growth all around the countryside. Yet we believe we know who the very character behind this veil of nature is – it is the spirit of the woods. That spirit of the Woods in Medieval England has been captured by many characters – as previously mentioned, Robin Goodfellow, yet we also have Jack in the Green, Oberon and the Fairies, including Puck from Shakespeare. We also have Robin Hood (or Robin of the Wood) and his 'Merry' Men. Again a long and deep folklore storyline with no real evidence to confirm if this was a real character or not. But I will attempt to make some small connections to Robin Goodfellow here... The name Robin comes from the Norman name Robert (which has earlier Germanic links), the word Robert means 'illumination, bright, to shine, enlightenment' – and we all know that both characters, Robin Goodfellow and Robin Hood were shining examples of true enlightenment. They were both called Robin, both dressed in green, they were Merry, they lived in the enchanted forests, and both gave gifts to the deserving and picked on the bad. Robin Hood was a master of disguise and Robin Goodfellow was a shape changer. So, is Robin Hood and Robin Goodfellow both based upon the same individual? I cannot say with any certainty, but everything seems to point this way. It was also interesting to read in Phyllis Siefker's book about Santa, that the Wildman was often termed 'Hood' or 'Wood' in Medieval times. I think my mind is made up on this, but others might require further evidence first, so this falls outside the scope of this book, but it is very interesting looking at this character in a different way.

It was very recent, March 4th, 2023 to be precise, that I stumbled upon something else that I don't think has ever been considered before. I had ventured off to Sutton Benger to obtain the above picture of one of the best-preserved Green Man Stone Carvings in the country. Sutton Benger's Church (All Saints Church) is a 12th Century church with a 15th century tower built upon it. It is a quaint church and well worth the visit as it has another 5 Green Men located on the outside lintel corners and depending upon the prevailing weather, they are each in different states of repair. Three of these are shown below.

The most weathered of the carvings.

All Saints Church, Sutton Benger.

Oak and acorn hidden Green Man.

Another wonderful carving, age uncertain.

So, you can imagine my surprise when faced with the following carving, which certainly looks very, very old, so I would say Medieval at least in age. What is this carving doing here alongside the Green Man carvings on the building lintels? I only noticed two other animal carvings, with most other carvings being Medieval human character faces. So, again, I ask myself, what is this doing here?

I think you can see the resemblance as much as I can. It's clearly an ape like figure. So, what is it doing here on a 12th Century Church? Is Sutton Benger the only Medieval Church with this type of representation? Can we make a link here with the Robin Goodfellow references to being a shapeshifter into an ape? As this only happened recently, given that I wanted to include some nice pictures of the Green Man for this book, I am left very, very confused, as I wasn't expecting this. I have seen some ape carvings before in some choir stalls, but that was a rarity I had thought. Now I am starting to think that there must be a lot more like this and this also helps point towards the Green Man being associated with these ape-like figures and I certainly wasn't expecting that to be confirmed by my visit to a Church!

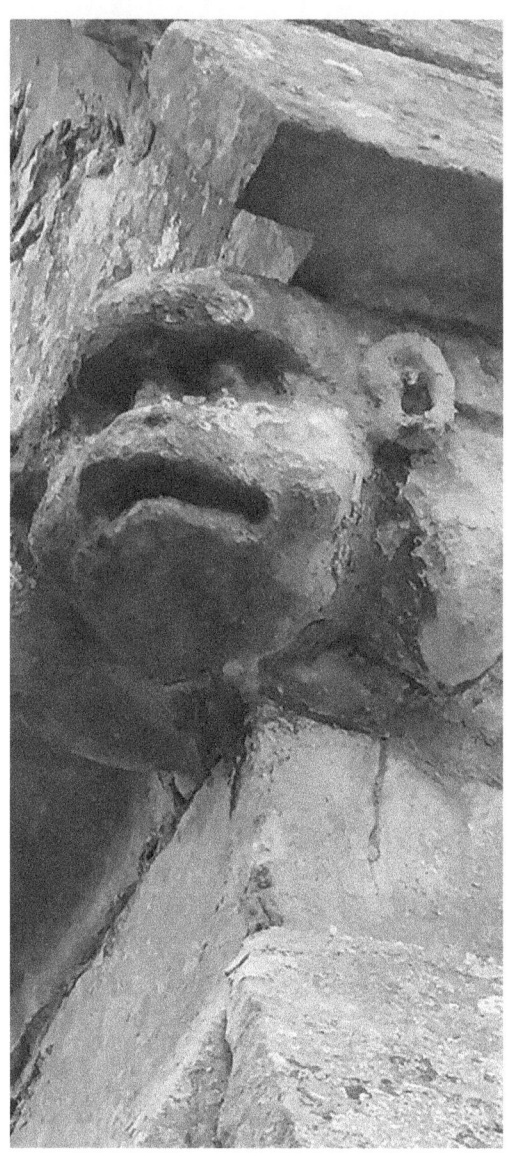

Sutton Benger. This carving should not have been here, yet it was.
Why was it added to a 12th. Century Church?

OTHER FOOD FOR THOUGHT

The following pages are my random thoughts of thinking outside of the box. None of this is based upon any factual evidence, but I feel it is worthy of noting for further consideration. It is this type of thought process that can lead to other connections being made (or not). Other people may decide to run with these or at least think more deeply about this, given the context from the previous chapters. As I say, keep an open mind!

ROBINS

The Humble Robin

I am sure that you will have seen plenty of Christmas cards with a Robin featured upon it. Should that be strange, given that other birds are not normally represented in the same festive way at this time of year? Well, the Robin has been voted on several occasions as the most popular bird in the United Kingdom, he is our favourite feathered friend! Why is this and why is he so popular at this time of year? Well, I am sure that you can appreciate some of the old wives' tales in that when you see a Robin, then you know that a lost one is close by. Again, there is a Christian belief in that the red breast of the Robin is from the crucifixion when the bird removed a thorn from the head of Jesus and the blood stained his front. These are all fascinating ideas, but one thing is for sure is that the name Robin is important. Not many birds are given human first names – apart from the Robin, the Jack-daw is possibly another one, yet it is the Robin that sticks out. Robins must have had some special affection many years ago in order to be granted a human name and that affection has transferred onto our popular Christmas cards. Why is this? I can imagine that years ago when people were ploughing the fields mid-winter that many Robins would drop down to the newly ploughed land and pick out worms. Farmers might have gained friendships with some that became tame, just like they do in many gardens to this date. There was something special about Robins, hence being given a human name. Now, I am not exactly sure when the Robin got his name but certainly this dates back to the 15th Century and gained the names Robin Redbreast, and even earlier dates names as Ruddock and Robinet. Robin, being short for Robert was a popular name at this time.

Did you also know that there is a national day of the Robin? Can you guess which day this is celebrated? Well, ironically, December 21st – i.e. the Winter Solstice! Incredible coincidence don't you think?

So, where am I leading with this? Could there have been a connection between Robin Goodfellow and Robin the bird? Was there a belief in the past that they are related or one of the same – given Robin was a shape shifter? One thing does remain from this medieval period is that they both attain the same name and were both popular and formed a special affection with humans at this time. Robin Goodfellow's connection to

Christmas might have been lost, but the bird Robin has certainly been welded into our modern-day festive traditions. I would like to think that there is some lost connection there, yet I have no way of proving this. Time will only tell if this is to be confirmed.

The Star of Bethlehem

Symbolism

When writing this chapter, it dawned on me that we sometimes put a star at the top of the Christmas Tree.... representing the star of Bethlehem. Now, could this in fact be a direct link to the brightest star in the sky that we also celebrate at this time - our sun? It is said that the bright star was seen in the east.... again, a possible link to the rising sun on the solstice.... in the east! Well, I cannot draw any conclusions from this.... but the symbolism is there, directly, indirectly or maybe purely coincidental? Also, it is noted that the Roman festival of Sol Invictus (December 25th) was a celebration for the birth of the 'Sun God' - Sol, and it was a Roman Emperor that gave rise to December 25th being celebrated as the birth of the 'Son of God'. So, Sun God and Son of God both celebrated on the same day seems a bit too much of a coincidence that we cannot ignore.

Also, the three wise men, if we are to accept the Christian narrative, came from Persia, India and Arabia. All of these areas are located to the East of Bethlehem. However, the star was seen in the East which led to the wise men travelling to Bethlehem… i.e. to the west. The star was not something that was noticed by the general population, so what was it to lead the wise men west? Astronomers understand there was a 'broom star' (comet) in 5 BC, based upon Chinese astronomical records, being the closest sizeable event that took place around this time. But there is little else to point to anything else taking place at that time, other than maybe, planetary alignments.... yet this makes little sense, as the wise men were led by a star in the east. And my understanding is that the star they were referring to was indeed our own sun and this is what they were celebrating – the winter Solstice.

The Christmas Tree

Toby, the author's Labrador sitting next to the Christmas Tree, December 2022.

We all understand the reason for the Christmas Tree being part of our festivities at this time of year. In the Middle Ages, our ancestors would 'Deck the hall with boughs of Holly' (this is from a Welsh 16th century melody). It was tradition to bring evergreen plants into the home at this time of year, as these were the only plants that stayed green at this cold time. Holly and ivy and evergreen branches were used to decorate homes with the idea that this will help goad in a good new harvest year. It was later in Victorian times, that we also introduced a Christmas tree to the home. It is also worthy of noting that we used to (and still do) place a fairy on the tree together with fairy lights. What amazing symbolism! The tree may not have Christian origins, yet we have absorbed this in to our modern day traditions. For me, this is a wonderful amalgamation of two very different mindsets, same as to how Santa developed over the years. Our traditions have changed over time and have morphed into what we know today, yet it is also very rewarding to know where they all came from and why they changed and now, we can even celebrate that some of these concepts are still very much alive today!

A Christmas Tree Fairy, with a difference.

Above you can see this Christmas Tree Fairy that I purchased to go on our tree last year. In essence, this fairy just caught my imagination as it appears like the fairy is sitting on an Orb. With the clear ball she is sitting on, lit from behind, it looks like a glowing orb, in the manner that we sometimes see them. This will be a real treasure to keep for many years to come… but I do hope other people, having seen this, might start producing something similar in the future.

The Son of God

This is where I am out of my depth, but I feel it could be a point of conversation. If we accept that Jesus lived, and I honestly believe something did indeed happen all those years ago which made the Roman Empire give up Pagan beliefs and transform to Christianity, then who was this incredible person that gave us teachings on how to be better people? Religion can be a dangerous subject to dig deep into, yet I feel given the bombshells that have been discussed earlier, then can we dare to go further? We understand that the Pagan green man was considered to be evil – and labelled as Satan – a goat footed horned beast, yet Satan's earliest descriptions were that he was a fallen angel and there was no reference to these later linked descriptions. And as we know to this date, the green man is never witnessed as looking like that fabled creature. My own personal experience is that they were humanoid and ape like, but that is just the tip of the ice burg in representing them. We know from Robin Goodfellow that he is a shapeshifter, the pamphlet from 1628 indicates that he can change his shape to anything desired – including large black dogs and even other people. So, could we dare to think that Robin Goodfellow could also be a religious figurehead too?

What do we know about Jesus? Well, he had the ability to heal people just by touching them, he could convert a small amount of food to feed a host of people. He had the ability to walk on water. He was a good judge of character and preached wise words. He arose from the dead following crucifixion.

Quite a tall order to follow? Agreed, yet some points we can find modern references to. We are aware that these forest people have healed people by using energy waves purely by their interactions with individuals (SOHA & SOIA encounters (Southern Oregon Habituation & Interaction Areas)). The wild man has the ability to use the power of the mind to either communicate or to read peoples thoughts, indeed, they have the ability to see into the future – and this is based upon my own experiences. They are magical and being shape changers then they had the ability to 'disappear' and then 'reappear' at a later stage. Walking on water would be no issue to them. Based upon my own experiences, then I am certain that they also have many more abilities which would constitute historical god like beings.

It also became apparent to me that as well as Christmas, the celebration of Easter also coincides with the Pagan celebrations of Spring and the re-birth of everything following the depths of winter. Indeed, the name Easter itself relates to the Anglo-Saxon Goddess 'Eostre' who was a God of Fertility and the God of Dawn and Light. So, as well as Christmas, other Christian celebrations have been blended over time in order to overlap the meanings and reasons for the celebrations at this time of year. With the Christian theme, the crucifixion of Christ, his dying and his resurrection on Easter Sunday is all about 're-birth' which coincides with the Pagan celebrations regarding the re-birth of nature during Spring. So, I have only briefly touched on this, as I am no academic for looking at the finer details…. but we can now see how some of our earliest celebrations have morphed into our current Christian traditions. That said, maybe, just maybe, both Pagan beliefs and Christian beliefs are both celebrating the same entity?

Where does this lead us? In all honesty, I have no idea. This is a huge subject and I have only scratched the surface. There is a much bigger picture to look at now. Were other historical magical Gods one of these beings? What I can say is that we are only now starting to accept and understand that we are not the only intelligent life force living on this planet, let alone the universe. Regarding the forest people, they are far more complex and have abilities far more advanced than us and as a result, they do deserve our absolute respect. Jokingly, you don't want to get on the 'naughty list' for these people and this is partly why Bigfoot hunters to date have not found their evidence, because trickery and deceit is not the way to make valuable contact with them.

Making Contact

As they say... Seeing is believing!

When I first started on my journey, to see if there was any validity to the wild man of the UK, I was totally out of my depth and relied upon the practices of the Bigfoot hunting community in North America. We know what was involved with that – knocking on trees at night-time, setting up trail cameras and use long recording sound equipment, vocalisations and searching for footprints and any tree features. So, over time I looked for similar things when researching in our UK woodlands. Overall, I have never done a wood knock or vocalisation during my research.... but I did always bow my head in the direction of the route into the forest I would take – out of respect for the forest people. This was in total contrast to other researchers, and I feel this simple thing was to benefit me in future years. Initially I was finding the woodland structures such as tree arches, sapling breaks, and tepee structures and then later at site 4, lots and lots of ground sticks. It is crazy to think that they were there already watching my activity at site 1.... yet I had no knowledge of this. They did leave me signs... the tree knocking stick, fresh tree breaks, some with, I believe, nail markings and then the tree that got pushed over along the forest path and near to where I had a trail camera set up. This was all great anecdotal evidence, yet it lacked that personal witness episode.

So, when we did our first night-time expedition, then all of those doubts were brushed aside. They were here and this first experience was recorded within that 29 minute recording I have added on Youtube – 'Proof that we have Bigfoot in the UK' and '9th. July, 2014 Sasquatch encounter – full 29 mins for analysis'. We did nothing to attract their attention.... but we had used technology surrounding us for recording any activity – as confirmed within the two recordings on Youtube. So, they found us and it freaked us both out and we left shortly after making that recording.... but, we came back! Over the weeks and months, we had all sorts of unusual activity and yet we made no effort to attract their attention.... and over time we learnt that the technology is not the best way forward. Trickery doesn't work as they are far smarter than we portrayed them to be (i.e. gigantopithecus etc.). So gradually, we dropped all forms of technology, and this is when amazing things started to happen. We had the 9 footer walk straight past us in pitch black darkness and due to not switching on any lights, then we were rewarded with a joyous feeling for three days – this is referred to as 'universal love'. It was amazing, truly amazing and we have no idea how they did this... but I started to understand that energy was key. We were seeing the orbs in the forest on a regular basis, so what were they? They had to be forms of energy. It was also at this time that things were happening at SOHA in America. This got me thinking... how can we make any sense of this? We were already getting much better contact with these forest beings; they had been getting closer and closer to us all the time. We had the smells – these changed from horrid smells at the beginning of our research to sweeter and much nicer smells. We have smelt honeysuckle, marzipan, strong pine smells amongst others. We had seen them as smoky figures, hot thermal signatures, red eye shine to name a few. So, it was time to think outside of the box.

The lessons learnt so far were to treat these forest people with respect – no trickery. Over time, build up some level of trust. Let them come to you as they will already know that you want to make contact. Talk to them, play music for them, just sit there in total darkness in their forest on their terms. If you are lucky then there is a chance that you will also make contact with them and maybe even have some level of mind speak with them and when that happens, be prepared to be blown away. If you can train your eyesight (it takes a while), and it didn't come easy for me, then you will be able to observe them in their interdimensional/cloaked form – which is their normal form. Once you can see them like this, then there is no 'un-seeing' them. A word of warning, the light levels need to be just right in order to achieve this and for me, dusk is the best time when the sun is not as bright etc. In time, what you will find is that they will await your arrival at the wood at the entrance point. You see, they already know when you are going to arrive and will wait for you.... now isn't that just incredible? This is why I refer to them as our forest friends because that is the best way that I can describe them from my own experience and hence, I believe, they got the name Robin Goodfellow all those years ago. I will underline that if you decide to do something that they don't like, then be warned, they can show their disdain. They fried my car's alternator on one occasion, and they gave me and Ricky an uneasy feeling when leaving the forest, together with a super loud heavy wood-knock just 30 feet behind us. Scary! <u>Learn from my lessons so that you don't repeat them!</u>

So, how do you get to the point that you see 300 orbs of glowing balls of light, or like shooting stars or fairy lights going off right in front of you? If you are looking for your own experiences and seeing an orb for yourself (although no guarantees – as they will only interact with individuals that they choose), then read on.

Energy. Energy... I stress energy. These beings fit outside of our current understanding. So... taking the enlightening words from a very great man:

"If you want to find the secrets of the universe, think in terms of energy, frequency and vibration."

Nikola Tesla was a genius, ahead of his time and I think what he said here was a clear reference to what I was dealing with. On a sub atomic level, even we are made of pure energy and yet all we see is the physical barriers that define us. So, should our definitions define all other life forms in the universe? Well, that would be very closed minded, so taking a look at Tesla's quote – if we are to unlock the secrets to the universe then we need to think about energy, frequency and vibration. Now, I am no physicist, but it came to me that I should try using Tesla's ideas in this subject area, and so this is exactly what happened. On 23rd. September, 2020 we tried out a theory and took a small Tesla Coil to the forest and see if anything would happen.

Here is the update I did for Project Zulu following this night:

23rd. of September, 2020

This night marked a major step forward in our contact with the forest people. We deployed for the first time the gadget for creating an electric field or arc at our normal spot. The reaction to this was crystal clear - all three present witnessed orbs flying around the device. If we counted them, then we would have lost count, but we can clearly say way in excess of 50. Most were attracted to the electric field we had produced and stayed close to this - at about a foot away up to a metre - but we also saw some others around us in the tree canopy. I witnessed the brightest one I have ever seen here - at about 7 feet above Ricky's head and shooting off right above us into the trees. Some were short lived and not so bright, others were a few seconds long and bright. The success of tonight was incredible. Normally we would see occasional orbs, but these were scattered all over the place - this time we only needed to look at one single point and the forest people came to that point. Some, all three of us saw, mostly Ricky and myself saw them at the same time and other times we each saw different ones at different times. It was truly amazing that we were able to have this level of visual interactions with them. Tonight history was made.... and going forward I am looking to increase the array of electrical fields to gain even better responses.'

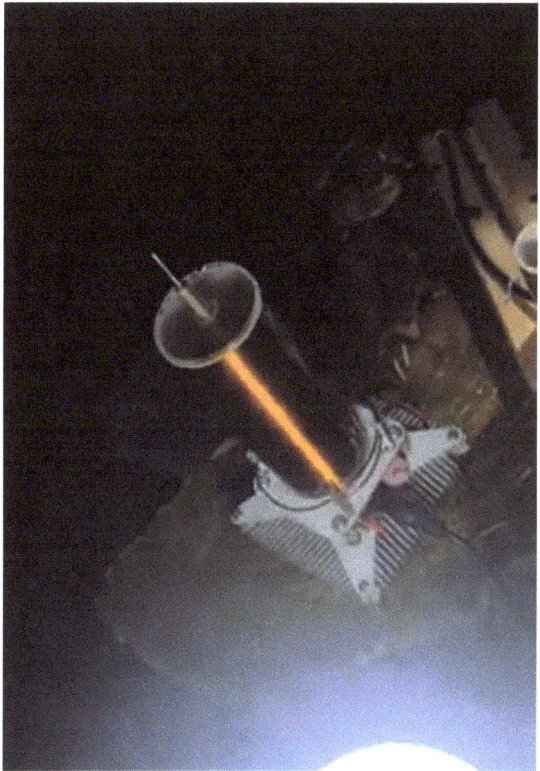

A tesla coil that we take out to the forest with us, deployed at site 6.

Well, it worked. It was only a small coil but it was having results. Small sparks and orbs were witnessed by the three people who were there that night. This was a result, so the idea was to build on this and now we take out a car battery, power inverter and an 18v Tesla Coil and the results speak for themselves. We were getting lots of orbs interacting with the unit and to the point of one night seeing over 300 orbs in close proximity to me in the space of a few hours. Now... I will refer to the fact that we are dealing with intelligent life forms here, so there is no guarantee that a clan will interact with it, like they do for us... but I am sure over time, patience will lead to rewards. Some nights we saw very few orbs, but now we see shimmering 7 to 9 foot figures just standing next to us and the Tesla device!

Now you know the secret to my success. I would ask that you are very careful and respectful when trying to follow this method, but this will certainly help turn the tables surrounding the whole 'woo' debate in the Bigfoot world. Rather than being labelled as pioneers in this field, we have been called wackos! It's now time for some to eat their words and apologise. Dare they?

So, in summary, build up a level of trust with a local clan. Show respect and spark up friendships. No trickery.... (I will add, the orbs stopped when I started filming), be on their terms – in their home habitat at night... this is beneficial in many ways as a lot of the orbs are faint – just like a fairy light! And if using the Tesla device, best to use it on moonless nights, or with limited moonlight. If you gain their trust, then you will witness some very bright orbs – just like shooting stars shooting through the forest canopy. Good luck and report back your findings, so the Bigfoot debate can open a new chapter.

There is one word of warning I have for those that have a lack of respect for these forest fellows... if you play with fire, then accept the fact that you will get burnt. I have reached this point based upon friendship with these people, don't think you can outsmart them... you can't. Be their friend and always show respect towards them.

This is a brave new world we are entering.

An Orb or not an Orb?

I see countless pictures of 'orbs' in Bigfoot discussion groups. These have mainly been taken at night time using a flash or an infra-red camera lens. They show singular dots of light or moving light sources around a room/woodland. Well, I am in no doubt ready to accept that these forest people can be in your home – as they live here at home with me too... but I have only ever seen one small orb at night in my home when I was half asleep in bed. I can safely say that the picture anomalies that we were recording in the forest at night were not orbs. Orbs you see as flashes of light. When I tried filming the orbs on the night that I witnessed 300 right in front of me, I tried filming... and then they all stopped showing themselves. So, knowing this, I decided to do my own bit of research and confirm one way or another what these light anomalies were. Were they insects, pollen grains or dust floating around? So, to replicate this, I used the dust bag from my hoover and shacked it outside in the dark and took a picture. The following is the result, and it confirms things:

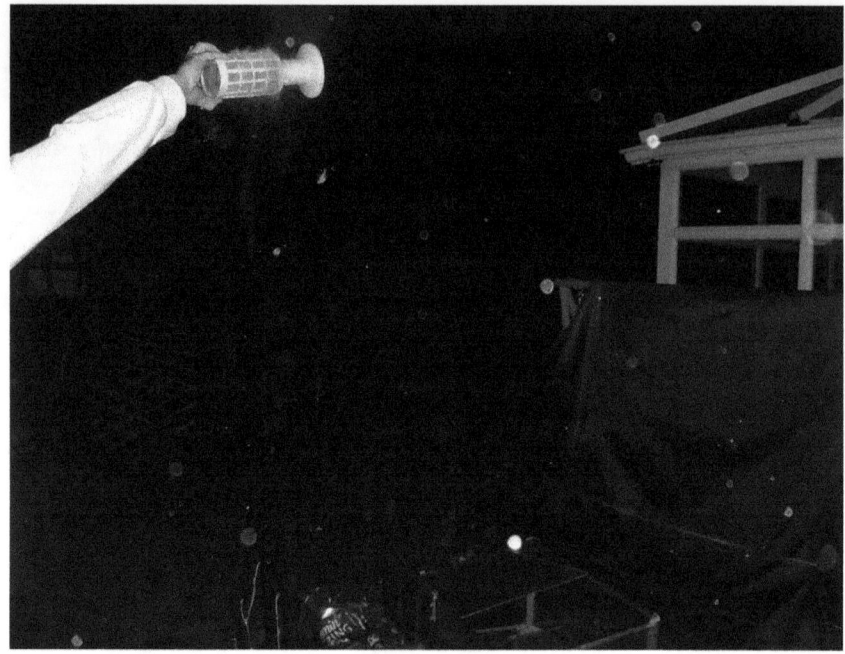

Orbs or just dust particles?

What else did I learn?

Through the years of my research and the mind-speak I have had, various things came to my notice, yet I have no way of proving any of it. So, please don't quote me on these, but feel that this will open the door to a more open mind about the new world that we now live in.

You will also notice from the Mad Pranks and Merry Jests of Robin Goodfellow that the forest people have the ability to shapeshift and one of the most common occurrences for this have been large black cats (and dogs in the case of the 1628 pamphlet). Many people have seen these large black cats and have no explanation for them. None have been shot – despite several attempts trying to track them down. What seems to be a trend is that the people who witness them usually end up having further experiences with the Xanue Forest People later in life. This was the case for both Gordon and Ricky and a few others that I have become aware of over time. Yet it doesn't end with black cats and dogs!

Thunderbirds, Nessie, the Moth-man and the likes – these are real, but just like Bigfoot, they are not confined to this dimension. They are energies that have broken through the fabric of the vibration that we occupy and are revealed for a short period. This might be induced by some natural occurrence, by cosmic distortions, or by their own abilities.

The Xanue are not the only apelike beings on this planet. There are many others around the world. Some are indeed missing links in the evolutionary world, and some are not Xanue, but are more aggressive in nature, yet they will all generally keep away from man.

We need to be more mindful of the damage that we are doing to our world. Please live in harmony with the planet. (Note: I always take out a bag with me and pick up any human litter I see in the Forest). They want us to get out into nature more and the more you learn about it, then the more you will respect it. Sometimes you can lose things that you can never replace. Also, please take care of yourself... your soul is very important to your future.

You are not alone in this vast universe. Many have visited you already and the time has come for you to understand this.

Artistic representation of the green man.

This is possibly the most important message I have picked up

Did you ever wonder why Christmas Eve felt so magical, and I mean truly magical, some years more than others? Was that pure coincidence? Well, my understanding is that it isn't, and this is truly, truly amazing. The forest people feel the love and emotion resonating from ourselves at this time of year… they are attracted to it and yes, it would not be unusual for them to pay you a visit on Christmas Eve and share their love, sometimes you can sense that electricity in the air when they do. The energy that we give off is spiritual and psychic and they pick up on this. So, in essence, you really are being visited by Father Christmas from time to time on this most special night.

Next time you put out a plate of bread and milk next to your Christmas Tree, then just remember this is you making a formal invitation and offering to them, and you are following an over 500 year old tradition!

My offering on the 24th. December, 2022 – including Hobgoblin ale and using a mini Tesla Coil.

Conclusions

When I say conclusions, I mean that there are no overall conclusions. This book is about the journey that I have taken over the last 10 years in search of the Wild Man in the United Kingdom. The only conclusion is that they are real, very real, they are no fairy tale! They are not consigned to medieval history books as something that could have existed all those years ago.... They are here living alongside us today. I remember asking the late Professor Sykes to come visit the forest near me, yet having travelled half way around the world looking for evidence, he couldn't travel just a few miles down the road to see for himself. It made me chuckle that I now know that there is a forest just outside Oxford where a clan of forest people live. So, within just a mile or so of the university where he was researching into the evidence for Bigfoot, they were literally right on his doorstep! How ironic. He has now left us, so he will never know this, which is a shame.

History, take notice here. Keep an open mind to what you are seeking. There have been major, major disagreements in the Bigfoot community about what type of creature Bigfoot is. Mainstream researchers stick to the concept that it must be a missing link and that those who have gone down the 'woo' path are complete idiots. I have had more than my fair share of ridicule over the years... even this year too, there is no real change. Yet having an open mind and showing a sense of respect for these forest people, then this book proves the results. I am hopeful that other researchers will follow suit and copy the techniques that I have used here and evidence things for themselves.

With regards to mind-speak. This is all personal. Some may achieve this, most people won't. To confirm the existence of the forest 'good fellows' here in the UK is one thing, to have mind speak with them to the point of knowing the name of the legendary Patty is another thing. To then know about some of the powers that these forest people have, is incredible. Then to find out that a very significant character from my childhood is actually based upon one of these forest folk, well, that just blew my mind! To know that Shakespeare and Charles Dickens were writing about Bigfoot has.... well, I am left speechless regarding that part. I have also touched upon some subjects that are dangerous and cannot be answered easily. Faith is a personal viewpoint and I respect people for their faith, and I do not want to step on people's toes regarding this. I have left the door open for people to ponder on, research further, evidence for themselves based upon the things I have said. History has a way of re-writing itself. Times change as do people's beliefs and feelings. Take a look at Santa to start with – how he has changed in the space of just 100 years, let alone 2000 years of history.

I wish I could have spent a lot more time writing this book, however, I don't get much free time outside of work. It was a rush to do this given every other avenue I had approached dare not touch this. Yet, I hope that this won't be the last book that I write on this subject and wish that I will be able to uncover even more amazing things about our forest friends. That all remains to be seen, but keeping everything crossed for now. If you have reached this far reading this book, then I thank you for your company sharing this journey with me. For my passing words, I will leave this up to our forest friends, who just wanted to remind me of something they once told me when I asked them regarding 'what is the meaning of life', which they then added an additional comment for me:-

'Does life have meaning for you?
If not, then you are the best person to create it.
You are the master of your own destiny.'

Special thanks must also go to my guide and guardian, Sol (Solomon). I first got to mind-speak with Sol back in 2017, together with Zach, but Sol has become my main contact ever since. Why Sol? Of all the names under the sun? Was this cryptic? Sol is a term for the 'Sun' from the Roman 'Sun God'. My journey was to take me to understand many things, but the main one being the significance of the winter 'SOLstice' and the 'SOL Invictus' being celebrated on December 25th.- Together with understanding the special character that visits us on this festive night! Was it just another coincidence… like all the others that came before it? Well, Enrith/Patty/Flower is yet another name that carries a hidden meaning… so I feel back in 2017, my fate had already been sealed. The world really is a magical place now!

Enrith, aka 'Flower', as drawn for me by Arfon Jones.
To my Forest Friends, I would like to thank you all for the
journey that I have had leading to writing this book.

Ricky's Testimonial

I started on this project around seven years ago myself. To begin with, I was like any other in the belief that it was all nonsense. What were the chances of these beings existing this close to home, here within the UK? However, I heard Paul out and decided to come along one afternoon to check out if what he was claiming held any weight.

I remember this afternoon vividly. The first thing I was immediately made aware of was the number of ground sticks in the area. At the time I was new to the tell-tale signs of Sasquatch activity, so like most, would have walked right past them completely unaware of the significance they held. Even with open eyes, at first, I still found these upended sticks difficult to spot despite there being so many in the surrounding area; you really do need to develop an eye for these, but once you have it, you have it!

I soon discovered that they were in total abundance around us. As many as the number of trees themselves. Perhaps this phenomenon was normal, they were everywhere - the result of tree felling or strong winds etc., however there were also as many parts of the forest void of any of these sticks. It didn't make sense to me.

As we continued on, more signs of activity were shown to me. A twisted tree and several tree arches in the brush. Ever sceptical, I tried to fathom out a natural way that these could have developed. Yes, perhaps a flexible sapling simply got caught and pinned down by a fallen branch from above, but why were there so many in the same area than anywhere else? I must say at the time, the more stuff I was made aware of, the more and more I began to feel a little on edge in the area. So, after a few hours we called time on the trip and headed home.

With excitement and my curiosity well and truly sparked, myself, and Paul, would continue to go out to the forest every other week keeping detailed recordings of our activities and findings. Many times, we would return with nothing to report, but almost equally there would be nights of exciting significance. Experimenting in different ways in the hopes to establish any form of contact with these illusive beings. We quickly discovered using technology was not really the best approach at achieving results. Trail cameras would fail and fully charged recorders would suddenly be drained off all battery power. That is not to say that we didn't have some success, as on several nights we managed to capture some solid activity, however we really did find that we achieved far, far more results when we ventured out without the gadgets. It was as if they knew.

In terms of material results over the years, we had managed to obtain audio evidence of movement around a set camera which included a wood knock before the footage immediately cut out, as well as visual footage of a Sasquatch walking in daylight before it seemingly disappearing into thin air. Using thermal imaging camera, I had once managed to catch a Sasquatch trekking through the brush, which I did draw a depiction of whilst fresh in the mind. This is shown earlier on in the book – 16th. July, 2015.

There have been many exciting, and very often emotional moments myself and Paul have each shared on this journey but if I were to have to pick a standout moment for me, it would always be a night back in May 2019, when we stood facing a 7ft tall being known to us now as 'Morah'. She stepped out from the undergrowth about 100ft away from where we were positioned. A very clear silhouette which faded and eventually vanished as we walked closer toward her in the dark. Paul received the name of her that night, which we later discovered translated from Hebrew into 'The Teacher'. That night was truly breath-taking and one of which I will never ever forget.

So where are we today? Well hopefully this book will help you understand just how far we have ventured down the rabbit hole. We have gone from seeking where these beings are to now asking, who these beings really are. And what Paul has personally uncovered holds a serious volume of credibility. There are far too many links and coincidences now not to regard this discovery with significant importance. My life has changed for the better now, knowing what I now know. I endorse this book and would strongly recommend you maintain an open mind and absorb all that is explained within it.

Ricky standing next to a very long ground stick – September, 2015.

Gordon's Testimonial

My first interest in Bigfoot started after I saw a newspaper article in 1967 when I was a teenager. It told of 2 cowboys who had caught a creature on a movie camera as it walked across a creek bed. The creature had turned to look at them – and the article included a picture – now that famous picture - of the 'look back' the creature gave. A picture now blazed on many tee shirts, hats and coffee mugs.

After that, there was little about Bigfoot until Sky TV started putting out more documentaries and the internet was also getting established. This led me to take up my interest in Bigfoot again and I started to catch up with what research was going on.

I started to follow a researcher in the USA who claimed to have contact with Bigfoot clans. He invited me over to see for myself and I was staggered to see what he said was true! I saw and connected with the Bigfoot Forest People – it felt a mixture of disbelief and destiny! I had amazing interaction and connection with them.

They said they wanted to connect with humans more and that I could help them do that when I returned to the UK. On my return, I went into the woods and met a clan – with a clan leader called Ison. I started to set up a Facebook Group based on their proper name – the Xanue (said as Janue).

It was then that Paul Glover and I got in touch with each other and we have managed to have some awesome experiences and interactions with the Xanue. Mainly, we seek out others who show the right attitude and qualities which the Xanue want to see in humans.

Gordon Dodds
Hook, Hampshire.

5th. August, 2022. Gordon with the Author on a night sit at one of Gordon's locations. Lots of orbs and shimmering figures were seen that night. Purely coincidental that bright blue shirts were worn by both.

Additional Pictures to ponder on

Here's a collection of pictures about the different types of woodland features that we have found during our 10 year search, which we honestly believe are associated with the forest people. Their meanings remain a mystery, but they certainly have a hidden meaning. One day, we hope we can start to read this forest language. There are also pictures of other things that you might find interesting, or worth pondering upon. For the sceptics that have reached this part of the book, then I would ask them to see the features that we have witnessed in the forests and go out and look for these themselves. Pick any large and well-established forest in the UK, or elsewhere in the world, and go look for yourself. If it is people behind this, then there must be a huge secretive society regularly planting these features up and down the county… however, at some point, just like I did, you will end up coming to different conclusions… as the evidence leads that way. Enjoy the journey!

13th. March, 2016. A forest message, but what?

A wider view of the 'V' shaped glyph associated with a ground-stick.

13 March 2016 – Glyph with two broken parallel placed branches associated with a ground-stick.

The interesting 'broken' small ground stick... This shows signs of possible pressure being used to push the stick into the ground and it broke in the process. 13th. March, 2016.

Bottom of Site 6: Tree arch created by the tree in the foreground being pushed over to hold the tree tip down. This arch was associated with two large ground sticks positioned on the right of this photo (out of sight). Two years later, another tree had been pushed over pinning the tip of the tree down again… together with a new ground stick.

Looking at the same structure, but from the opposite side.

The ground-stick placed next to the tree arch at Site 6.

Just another 6 foot tall ground stick in site 1 - 13th. March, 2016.

Site 6 – Summer 2022. Our night time location, but the forest people had left us a huge branched ground stick up against the nearest tree.
Note: The branch is broadleaved and the trees around our spot are all Fir trees.

The author standing next to a long ground stick – yes, a stick that has been stuck in the ground! Why on Earth would a human do this? September 8th. 2015.

Random star shaped stick structure. Not normal dead fall from trees, one branch was a new addition pointing to repeated use. At the start of my research, I was led to believe that the number of branches here represented the number of individuals in the clan, with the recent addition meaning a new member added. I am uncertain about this myself, but it is a clear marker.

The Author standing next to a new ground stick in Site 4. June 9, 2015.

Ground sticks can be small as well as large. June 16th, 2015. Site 4. You can see how fresh this one is.

Just confirming that these ground-sticks are pushed into the ground. June 16th. 2015.

It didn't matter what the time of year was, ground-sticks would appear throughout the year. November 27th, 2015.

A very visible small ground stick that had been made from a split tree branch. 14th. December, 2015.

Sapling breaks at Site 1. These are at about waist height and above. This example isn't common as it does show evidence of deer antler damage towards it's base. You can spot the deer damage trees as there is usually more severe rubbing on the bark. The issue is that you see deer damaged saplings throughout the forest, although not common, yet here at site 1 there are hundreds of sapling breaks with no associated deer damage, but do have nail like markings instead.

2015. Thinking outside the box. How best to find long hair in the forest? Well, how about letting the Forest birds do that job for you. So, we were collecting old bird nests where possible and checking for any interesting hair samples. This turned out to be a waste of time, but the theory this was sound.

Inside the nest like structure. Very deceptive from the outside.

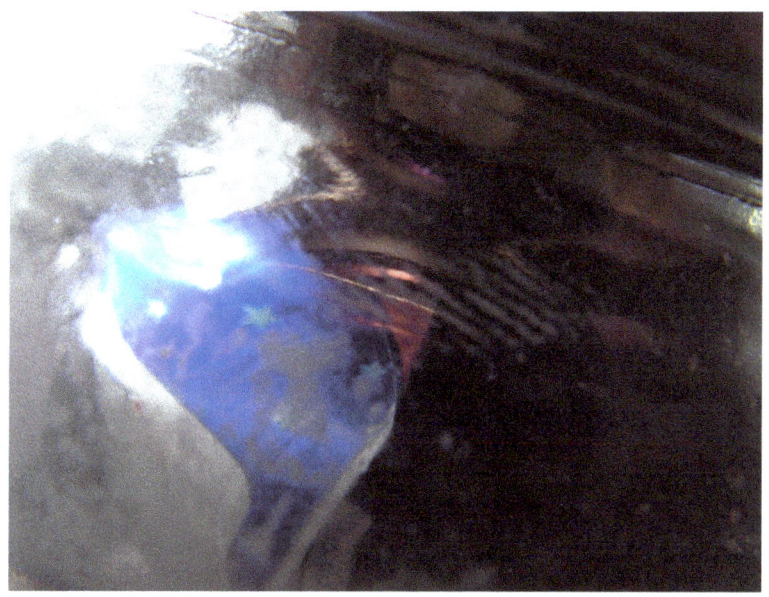

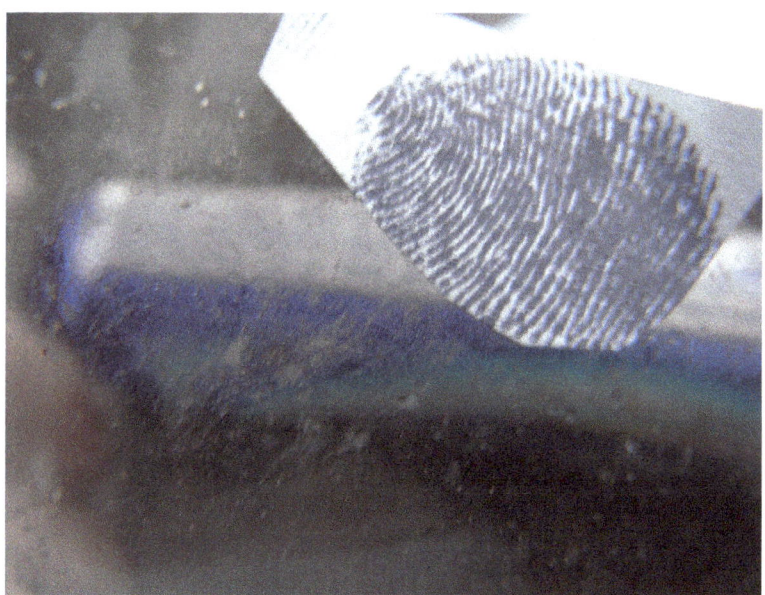

More dermal prints retrieved, notice the thickness and spacing between the dermal ridges. July 2015.

Other examples of the sapling breaks with associated scratch markings. March 5th. 2015.

Authors nail next to one of the features, for comparison.
Experiments on fresh branches using nails confirmed similar results.

Yet another sapling break with associated nail scratch like markings. Testing confirmed that human fingers can create similar scratches on the bark of a sapling, but some of these are far bigger.

Ricky next to a thick ground stick – Site 4, 27th. November, 2015.

This is a feature that you come across from time to time, a tree pushed over across a forest path. It would have taken considerable force to push this particular tree over…. the reason here, we believe, is due to the recent forestry work adjacent to this path. Site 5, Summer 2015.

Example of trail camera images captured at site 1, November, 2014.

Daytime image from a trail camera at site 1. November, 2014.

Gloucester Minster, taken on the author's visit in October 2020, hunting the Wildman. This was a carving within the Choir Stalls, understood to have been created around 1370. No mistaking what this is meant to represent.

Close up of the amazing Green Man of Sutton Berwick. You can see some of the damage to the missing leaves on the left side of the face. It is interesting that they still classed the Green Man as a person, with a human like face, yet spouting out vegetation from his mouth.

Looking at the carving from below focusing on the leaves coming from the mouth.

View from the side of the Green Man of Sutton Berwick. Notice the bird pecking at fruit, confirming a close relationship he has with nature.

Old Coins

These are some of the interesting Green Man/Bigfoot Medieval Coins that I have obtained over the years. These all come from Germany. The moment I came across the images of a hairy man carrying a candle, I automatically made the connection. I am certain this is exactly what they were depicting from the encounters people had in the 1500's. History has a funny way of trying to explain things that were unknown – Shakespeare did the same. Coins are a great way of understanding what has been going on at the time…. Just take a look at our own modern coinage, you see various examples that symbolise memorable events – Olympic & Commonwealth games, Monarch & Historical Anniversaries, Famous characters – past and present e.g. Darwin and Lady Diana, Brexit even. Some coins you will look at and by their image you would not make any connection to what they represent, but for the people of the time, this made absolute sense. For example, the 1978 50 pence coin with 9 clasped hands rotated around the coin. To anyone finding the coin will not understand what this represents, but knowing the significance of the year, 1978, the UK had assession to the EEC, then you can start to appreciate what it stands for. Likewise, these coins from the 1500's represent some features that wouldn't ordinarily make any sense, a hairy man carrying a candle? But knowing what we know now, we can easily recognise what this is referring to.

The association of these hairy wild men with lights cannot be overlooked. In the Middle Ages, they knew they were associated and today, we can now confirm that too, through our own investigations. Folklore was right again!

A lovely example of a hairy Wildman carrying a heavy branch (for wood-knocks?) and a candle.

Just like the lights, the tree arrangements hadn't gone un-noticed by Medieval people.

Hairy man holding a candle – 1587.

Hairy man carrying a candle with an uprooted tree – emphasising their strength – 1572.

A smaller version that has been well handled during the hundreds of years of it's existence.

The author's most expensive coin. Here you can see the representation of the Wildman closely resembling the character appearing in a Dickens Christmas Carol. One of the earliest representations of Father Christmas maybe?

Other Green Santa's

These come from the author's collection.

Christmas card from 1912 with a green Santa, together with a red hat!

Another green Santa, circa 1910.

Edwardian period hand painted photograph, with a green painted Father Christmas.

Time to Ponder…..

I hope this book has made you stop and think twice about certain things. I have had a long time to digest these findings based upon my own research and experience, and I accept this as fact, not fiction now. But, for those sceptics that are still having difficulties accepting these findings, then I have drawn up a list of things that most Bigfoot researchers are fully aware of or can easily relate to or find references for. Thinking about all of this, does it all add up now?

- Folklore or North American Indian stories about the supernatural abilities?
- How could such beings be present in densely populated United Kingdom?
- Why no physical evidence to date? A body or any other anatomical part?
- Why no fossil evidence for them?
- Orbs?
- Why electrical items sometimes fail?
- Claims that these beings can disappear in a flash of light?
- The Patterson Gimlin film of a potential second Sasquatch and a flash of light?
- Some footprint trails coming to a dead end, even in the snow?
- Rotting meat smells (or others) yet no physical presence?
- Infrasound or zapping?
- Glowing red eyes at night?
- People say that they have shot at them, but this did not affect them.
- All these large black cat sightings but no concrete evidence for them?
- People reputing that they have been healed at SOHA/SOIA?
- The TV being switched on by itself at 2:22 am?
- Mind-speak?
- Feelings of being watched or feelings of dread at some locations?

If after considering all of this and you still need further confirmation, then I have given away the secrets to me obtaining my own concrete evidence... I would ask you to follow suit, but in the full knowledge this will always be on their terms. When you see that first orb close to you, then you know that you are dealing with something exceptional. As a result, always show your respect to them and then you can look forward to more meaningful contact.

One last closing comment: Ancient Woodlands are those that have persisted in the same spot since the 1600's (and 1750 in Scotland). Only 2.5% of the UK is covered by these ancient woodlands, around 609,990 hectares. One fact I recently found out was that our woodlands are home to 50% of the world's collection of bluebells, and you will understand just how beautiful those bluebell woods look in April. Our woodlands are truly remarkable places, and we are only now starting to understand their importance and that it isn't just carbon-based life that exists in these magical places. I hope then, that with knowledge comes greater responsibility. We need to help preserve and expand our forests, for future generations and more importantly, for the natural world and state of the planet. If you have read this book, then please accept my thanks and understand that these very words are typed upon the remains of once majestic trees and I would ask that you in turn, with my absolute thanks, can take the time to plant a tree in return. Thank you.

BIGFOOT: IT'S A FAIRY TALE...

www.ingramcontent.com/pod-product-compliance
Lightning Source LLC
Chambersburg PA
CBHW040250170426
43191CB00018B/2367